PRINCIPLES OF DAIRY MANAGEMENT

Dr. P. Venkateshwara Rao (b. 1956–) has a doctorate degree in Veterinary Science. He has extensive teaching and research experience in a number of academic and research institutions. Dr. Rao is the author of three books and over 16 articles in reputed journals in India. He has participated in many international and national level seminars. He also provides guest lectures in various professional institutions.

PRINCIPLES OF DAIRY MANAGEMENT

by
P. Venkateshwara Rao

BIOTECH
2023
Biotech Books

ISBN10 81-7622-199-6
ISBN13 978-81-7622-199-3

Published by : **BIOTECH BOOKS**
4762-63/23, Ansari Road, Darya Ganj,
New Delhi - 110002
Phones : 011 4300 3222
Email : biotechbooks@yahoo.co.in

PRINTED IN INDIA

Preface

Dairying is the business of producing milk, butter, cheese and other milk products, and entails an industry which is continually growing the world over, with an annual turnover of nearly $500 billion. With mechanisation pervading nearly every arena of the agricultural process, dairy management has had to incorporate the principles and techniques which define the latest technology with age-old practices, leading to a pertinent amalgamation of the old with the new, an essential feature of the dairy farm endeavour. With the increasing insistence on health and safety checks, the relevance of these principles of dairy management has undertaken a whole new dimension altogether.

This text aims to acquaint dairy farmers with the most essential and relevant principles of the science of dairy management, taking care to thoroughly entrench them in the basic concepts, practices and techniques which give shape to the modern process of the science. In addition to delineating the current trends and developments in the dairying field, the text takes care to elaborate upon the whole industry, its revenues, prospects and challenges, with due consideration to the rapidly globalising dairy sector itself. Further, critical perspectives have been incorporated, which provide fresh insights into the ways in which dairy management is rapidly assuming new dimensions.

It is hoped that the erudition and insight of the book serves well for the readers.

P. Venkateshwara Rao

Contents

1

Introduction to Dairy Production

A dairy is a facility for the extraction and processing of animal milk—mostly from goats or cows, but also from buffalo, sheep, horses, or camels —for human consumption.

Terminology differs slightly between countries. In particular, in the U.S. a *dairy* can also be the facility that processes and distributes the milk or the store that sells dairy products, and in New Zealand English a *dairy* means a corner shop, or Superette—and *dairy factory* is the term for what is elsewhere a *dairy*.

As an adjective, the word *dairy* describes milk-based products, derivatives and processes, for emple dairy cattle, dairy goat. A dairy farm produces milk and a dairy factory processes it into a variety of dairy products.

Historical Outlook

Milk-producing animals have been domesticated for thousands of years. Initially they were part of the subsistence farming that nomads engaged in. As the community moved about the country so did their animals accompany them. Protecting and feeding the animals were a big part of the symbiotic relationship between the animal and the herder.

In the more recent past, people in agricultural societies owned dairy animals that they milked for domestic or local (village) consumption, a typical example of a cottage industry.

The animals might serve multiple purposes (for example, as a draught animal for pulling a plough as a youngster and at the end of its useful life as meat). In this case the animals were normally milked by hand and the herd size was quite small so that all of the animals could be milked in less than an hour—about 10 per milker.

With industrialisation and urbanisation the supply of milk became a commercial industry with specialised breeds of cow being developed for dairy, as distinct from beef or draught animals. Initially more people were employed as milkers but it soon turned to mechanisation with machines designed to do the milking.

As herd numbers increased so did the problems of animal health. In New Zealand two approaches to this problem have been used. The first was improved veterinary medicines (and the government regulation of the medicines) that the farmer could use. The other was the creation of *veterinary clubs* where groups of farmers would employ a veterinarian (vet) full-time and share those services throughout the year. It was in the vet's interest to keep the animals healthy and reduce the number of calls from farmers, rather than to ensure that the farmer needed to call for service and pay regularly.

Most dairy farmers milk their cows with absolute regularity at a minimum of twice a day, with some high-producing herds milking up to four times a day to lessen the weight of large volumes of milk in the udder of the cow. This daily milking routine goes on for about 300 to 320 days per year that the cow stays in milk. Some small herds are milked once a day for about the last 20 days of the production cycle but this is not usual for large herds. If a cow is left unmilked just once she is likely to reduce milk-production almost immediately and the rest of the season may see her *dried off* (giving no milk) and still consuming feed for no production. However, once-a-day milking is now being practised more widely in New Zealand for profit and lifestyle reasons. This is effective because the fall in milk yield is at least partially offset by labour and cost savings from milking once per day.

This compares to some intensive farm systems in the United States that milk three or more times per day due to higher milk yields per cow and lower marginal labor costs.

Farmers who are contracted to supply liquid milk for human consumption often have to manage their herd so that the contracted number of cows are in milk the year round, or the required minimum milk output is maintained. This is done by mating cows outside their natural mating time so that the period when each cow in the herd is giving maximum production is in rotation throughout the year.

Northern hemisphere farmers who keep cows in barns almost all the year usually manage their herds to give continuous production of milk so that they get paid all year round. In the southern hemisphere the cooperative dairying systems allow for two months on no productivity because their systems are designed to take advantage of maximum grass and milk production in the spring and because the milk processing plants pay bonuses in the dry (winter) season to carry the farmers through the mid-winter break from milking. It also means that cows have a rest from milk production when they are most heavily pregnant. Some year-round milk farms are penalised financially for over-production at any time in the year by being unable to sell their overproduction at current prices. Artificial insemination (AI) is common in all high-production herds.

Types of Milk Products

Cream and Butter

Today, milk is separated by large machines in bulk into cream and skim milk. The cream is processed to produce various consumer products, depending on its thickness, its suitability for culinary uses and consumer demand, which differs from place to place and country to country.

Some cream is dried and powdered, some is condensed (by evaporation) mixed with varying amounts of sugar and canned. Most cream from New Zealand and Australian factories is made into butter. This is done by churning the cream until the fat globules coagulate and form a monolithic mass. This butter mass is washed and, sometimes, salted to improve keeping qualities. The residual buttermilk goes on to further processing. The butter is packaged (25 to 50 kg boxes) and chilled for storage and sale. At a later stage these packages are broken down into home-consumption sized packs. Butter sells for about US$3200 a tonne on the international market in 2007 (an unusual high).

Skim Milk

The product left after the cream is removed is called skim, or skimmed, milk. Reacting skim milk with rennet or with an acid makes casein curds from the milk solids in skim milk, with whey as a residual. To make a consumable liquid a portion of cream is returned to the skim milk to make *low fat milk* (semi-skimmed) for human consumption. By varying the amount of cream returned, producers can make a variety of low-fat milks to suit their local market. Other products, such as calcium, vitamin D, and flavouring, are also added to appeal to consumers.

Casein

Casein is the predominant phosphoprotein found in fresh milk. It has a very wide range of uses from being a filler for human foods, such as in ice cream, to the manufacture of products such as fabric, adhesives, and plastics.

Cheese

Cheese is another product made from milk. Whole milk is reacted to form curds that can be compressed, processed and stored to form cheese. In countries where milk is legally allowed to be processed without pasteurisation a wide range of cheeses can be made using the bacteria naturally in the milk. In most other countries, the range of cheeses is smaller and the use of artificial cheese curing is greater. Whey is also the byproduct of this process.

Cheese has historically been an important way of "storing" milk over the year, and carrying over its nutritional value between prosperous years and fallow ones. It is a food product that, with bread and beer, dates back to prehistory in Middle Eastern and European cultures, and like them is subject to innumerable variety and local specificity. Although nowhere near as big as the market for cow's milk cheese, a considerable amount of cheese is made commercially from other milks, especially goat and sheep.

Whey

In earlier times whey was considered to be a waste product and it was, mostly, fed to pigs as a convenient means of disposal. Beginning about 1950, and mostly since about 1980, lactose and many other products, mainly food additives, are made from both casein and cheese whey.

Yogurt

Yoghurt (or yogurt) making is a process similar to cheese making, only the process is arrested before the curd becomes very hard.

Milk powders

Milk is also processed by various drying processes into powders. Whole milk and skim-milk powders for human and animal consumption and buttermilk (the residue from butter-making) powder is used for animal food. The main difference between production of powders for human or for animal consumption is in the protection of the process and the product from contamination. Some people drink milk reconstituted from powdered milk, because milk is about 88% water and it is much cheaper to transport the dried product. Dried skim milk powder is worth about US$5300 a tonne (mid-2007 prices) on the international market.

Other milk products

Kumis is produced commercially in Central Asia. Although it

is traditionally made from mare's milk, modern industrial variants may use cow's milk instead.

Transport of Milk

Historically, the milking and the processing took place in the same place: on a dairy farm. Later, cream was separated from the milk by machine, on the farm, and the cream was transported to a factory for buttermaking. The skim milk was fed to pigs. This allowed for the high cost of transport (taking the smallest volume high-value product), primitive trucks and the poor quality of roads. Only farms close to factories could afford to take whole milk, which was essential for cheesemaking in industrial quantities, to them. The development of refrigeration and better road transport, in the late 1950s, has meant that most farmers milk their cows and only temporarily store the milk in large refrigerated bulk tanks, whence it is later transported by truck to central processing facilities.

Milking Machines

Milking machines are used to harvest milk from cows when manual milking becomes inefficient or labour intensive. The milking unit is the portion of a milking machine for removing milk from an udder. It is made up of a claw, four teatcups,(Shells and rubber liners) long milk tube, long pulsation tube, and a pulsator. The claw is a assembly that connects the short pulse tubes and short milk tubes from the teatcups to the long pulse tube and long milk tube.(Cluster assembly) Claws are commonly made of stainless steel or plastic or both. Teatcups are composed of a rigid outer shell (stainless steel or plastic) that holds a soft inner liner or *inflation*. Transparent sections in the shell may allow viewing of liner collapse and milk flow. The annular space between the shell and liner is called the pulse chamber.

Milking machines work in a way that is different from hand milking or calf suckling. Continuous vacuum is applied inside the soft liner to massage milk from the teat by creating a pressure difference across the teat canal (or opening at the end of the teat). Vacuum also helps keep the machine attached to the cow. The vacuum applied to the teat causes congestion of teat tissues (accumulation of blood and other fluids). Atmospheric air is admitted into the pulsation chamber about once per second (the pulsation rate) to allow the liner to collapse around the end of teat and relieve congestion in the teat tissue. The ratio of the time that the liner is open (milking phase) and closed (rest phase) is called the pulsation ratio.

The four streams of milk from the teatcups are usually combined in the claw and transported to the milkline, or the collection bucket (usually sized to the output of one cow) in a single milk hose. Milk is then transported (manually in buckets) or with a combination of airflow and mechanical pump to a central storage vat or bulk tank. Milk is refrigerated on the farm in most countries either by passing through a heat-exchanger or in the bulk tank, or both.

Milking machines keep the milk enclosed and safe from external contamination. The interior 'milk contact' surfaces of the machine are kept clean by a manual or automated washing procedures implemented after milking is completed. Milk contact surfaces must comply with regulations requiring food-grade materials (typically stainless steel and special plastics and rubber compounds) and are easily cleaned.

Most milking machines are powered by electricity but, in case of electrical failure, there can be an alternative means of motive power, often an internal combustion engine, for the vacuum and milk pumps. Milk cows cannot tolerate delays in scheduled milking without serious milk production reductions.

Milking Shed Layouts

Bail-style sheds— This type of milking facility was the first development, after open-paddock milking, for many farmers. The building was a long, narrow, *lean-to* shed that was open along one long side. The cows were held in a yard at the open side and when they were about to be milked they were positioned in one of the bails (stalls). Usually the cows were restrained in the bail with a breech chain and a rope to restrain the outer back leg. The cow could not move about excessively and the milker could expect not to be kicked or trampled while sitting on a (three-legged) stool and milking into a bucket. When each cow was finished she backed out into the yard again.

As herd sizes increased a door was set into the front of each bail so that when the milking was done for any cow the milker could, after undoing the leg-rope and with a remote link, open the door and allow her to exit to the pasture. The door was closed, the next cow walked into the bail and was secured. When milking machines were introduced bails were set in pairs so that a cow was being milked in one paired bail while the other could be prepared for milking. When one was finished the machine's cups are swapped to the other cow. This is the same as for *Swingover Milking Parlours* as described below except that the cups are loaded on the udder from the side. As herd numbers increased it was easier to double-up the cup-sets and milk both cows simultaneously than to increase the number of bails. About 50 cows an hour can be milked in a shed with 8 bales by one person.

Herringbone Milking Parlours— In herringbone milking sheds, or parlours, cows enter, in single file, and line up almost perpendicular to the central aisle of the milking parlour on both sides of a central pit in which the milker works (you can visualise a fishbone with the ribs representing the cows and the spine being the milker's working area; the cows face outward). After washing the udder and teats the cups of the milking machine are applied to the cows, from the rear of their hind legs, on both sides of the working area. Large herringbone sheds can milk up to 600 cows efficiently with two people.

Swingover Milking Parlours— Swingover parlours are the same as herringbone parlours except they have only one set of milking cups to be shared between the two rows of cows, as one side is being milked the cows on the other side are moved out and replaced with unmilked ones. The advantage of this system is that it is less costly to equip, however it operates at slightly better than half-speed and one would not normally try to milk more than about 100 cows with one person.

Rotary Milking sheds— Rotary milking sheds consist of a turntable with about 12 to 100 individual stalls for cows around the outer edge. A "good" rotary will be operated with 24–32 stalls by one (two) milkers. The turntable is turned by an electric-motor drive at a rate that one turn is the time for a cow to be milked completely. As an empty stall passes the entrance a cow steps on, facing the centre, and rotates with the turntable. The next cow moves into the next vacant stall and so on. The operator, or milker, cleans the teats, attaches the cups and does any other feeding or whatever husbanding operations that are necessary. Cows are milked as the platform rotates. The milker, or an automatic device, removes the milking machine cups and the cow backs out and leaves at an exit just before the entrance. The rotary system is capable of milking very large herds—over a thousand cows.

Automatic Milking sheds— Automatic milking or 'robotic milking' sheds can be seen in many European countries. Current automatic milking sheds use the voluntary milking (VM) method. These allow the cows to voluntarily present themselves for milking at any time of the day or night, although repeat visits may be limited by the farmer through computer software. A robot arm is used to clean teats and apply milking equipment, while automated gates direct cow traffic, eliminating the need for the farmer to be present during the process. The entire process is computer controlled.

Supplementary accessories in sheds— Farmers soon realised that a milking shed was a good place to feed cows supplementary foods that overcame local dietary deficiencies or added to the cows' wellbeing and production. Each bail might have a box into which such feed is delivered as the cow arrives so that she is eating while being milked. A computer can read the eartag of each beast to ration the correct individual supplement.

The holding yard at the entrance of the shed is important as a means of keeping cows moving into the shed. Most yards have a powered gate that ensures that the cows are kept close to the shed. Water is a vital commodity on a dairy farm: cows drink about 20 gallons (80 litres) a day, sheds need water to cool and clean them. Pumps and reservoirs are common at milking facilities.

Temporary Milk Storage

Milk coming from the cow is transported to a nearby storage vessel by the airflow leaking around the cups on the cow or by a special "air inlet" (5-10 l/min free air) in the claw. From there it is pumped by a mechanical pump and cooled by a heat exchanger. The milk is then stored in a large vat, or bulk tank, which is usually refrigerated until collection for processing.

Waste Disposal

In countries where cows are grazed outside year-round there is little waste disposal to deal with. The most concentrated waste is at the milking shed where the animal waste is liquefied (during the water-washing process) and allowed to flow by gravity, or pumped, into composting ponds with anaerobic bacteria to consume the solids. The processed water and nutrients are then pumped back onto the pasture as irrigation and fertilizer. Surplus animals are slaughtered for processed meat and other rendered products.

In the associated milk processing factories most of the waste is washing water that is treated, usually by composting, and returned to waterways. This is much different from half a century ago when the main products were butter, cheese and casein, and the rest of the milk had to be disposed of as waste (sometimes as animal feed).

In areas where cows are housed all year round the waste problem is difficult because of the amount of feed that is bought in and the amount of bedding material that also has to be removed and composted. The size of the problem can be understood by standing downwind of the barns where such dairying goes on.

In many cases modern farms have very large quantities of milk to be transported to a factory for processing. If anything goes wrong with the milking, transport or processing facilities it can be a major disaster trying to dispose of enormous quantities of milk. If a road tanker overturns on a road the rescue crew is looking at accommodating the spill of 10 to 20 thousand gallons of milk (45 to 90 thousand litres) without allowing any into the waterways. A derailed rail tanker-train may involve 10 times that amount. Without refrigeration, milk is a fragile commodity and it is very damaging to the environment in its raw state. A widespread electrical power blackout is another disaster for the dairy industry because both milking and processing facilities are affected.

In dairy-intensive areas the simplest way of disposing of large quantities of milk has been to dig a large hole in the ground and allow the clay to filter the milk solids as it soaks away. This is not very satisfactory.

Common Diseases

Mastitis can also be a common disease found in milk which cause it to go off very quickly and has a horrible sour taste.

- Leptospirosis is one of the most common debilitating diseases of milkers, made somewhat worse since the introduction of herringbone sheds because of unavoidable direct contact with bovine urine
- Cowpox is one of the helpful diseases; it is barely harmful to humans and tends to inoculate them against other poxes such as chickenpox
- Tuberculosis (TB) is able to be transmitted from cattle mainly via milk products that are · unpasteurised and many dairy-producing families consume milk that way. In the important dairy exporting countries TB has been eradicated from herds by testing for the disease and culling suspected animals

— Brucellosis is a bacterial disease transmitted to humans by dairy products and direct animal contact. In the important dairy exporting countries Brucellosis has been eradicated from herds by testing for the disease and culling suspected animals
— Listeria is a bacterial disease associated with unpasteurised milk and can affect some cheeses made in traditional ways. Careful observance of the traditional cheesemaking methods achieves reasonable protection for the consumer.
— Johne's Disease (pronounced "yo-knees") is a contagious, chronic and usually fatal infection in ruminants caused by a bacterium named Mycobacterium avium subspecies paratuberculosis (M. paratuberculosis). The bacteria is present in retail milk and is believed by some researchers to be the primary cause of Crohn's disease in humans. This disease is not known to infect animals in Australia and New Zealand.

References

Jay, J.M., *Modern Food Microbiology,* Fourth Edition, New York: Chapman & Hall, 1992.

Potter, N.N., J. H. Hotchkiss, *Food Science,* Fifth Edition, New York: Chapman & Hall, 1995.

Swasigood, H.E., "Characteristics of Edible Fluids of Animal Origin: Milk", In *Food Chemistry,* Second Edition, Revised and Expanded, O.R. Fennema, Ed. New York: Marcell Dekker, Inc., 1985.

2

Organisation of Dairy Plant

Milk Collection

Many different patterns of milk collection exist in different countries. The pattern taken into consideration here is one whereby all milk is delivered by the producers to a collection and/or cooling centre. The collection centres belong to the same dairy organisation as the dairy processing plant. The milk is weighed-in and bulked in cans at the collection centre. The milk is transported to the processing plant only once a day and evening milk can therefore be accepted in collection centres with cooling facilities only. Farmer accounts are maintained by the accounts section at the processing dairy. The recording system worked out below should be adjusted for other patterns of milk collection.

To facilitate the recording and accounting of milk supplies each producer should be given a code. This code should have three elements:

- A code for the collection route;
- a code for the collection centre. This code preceded by the code for the route identifies each collection centre;
- a code for each producer supplying to a particular collection centre.

This code would also be useful in countries where a person can freely change his name. It also avoids confusion when a herdsboy or any other person supplies milk on behalf of a producer and gives his own name instead of the producer's name.

A register of producers should be kept at each collection centre for the producers supplying at that centre and one should be kept at the accounts office for all producers in the milkshed supplying to the processing plant. This register should show for each producer: his code number, his full name and any other particulars used in the country to identify a person.

The basic document for quantity control as well as for maintaining suppliers accounts is the Milk Receipt. This document should be maintained in triplicate by the milk buyer in each collection centre. The second copy should be perforated in such a way that each receipt can be torn off to be given to individual suppliers. The first copy should be perforated so that the whole page can be torn out of the book while the third copy should not be perforated. The first and third copy should have space at the bottom for recording the total of the page, the quantity brought forward and the cumulative quantity. Each individual receipt should be numbered.

When the milk buyer at the collection centre receives milk from a producer he should carry out the usual tests and when the milk is accepted he should weigh it and record the quantity on a Milk Receipt. He should further fill in the date and the farmers code, he should also make a mark in the space after AM or PM to indicate whether it was morning or evening milk. The milk buyer should then sign the Milk Receipt, tear off the second copy and give it to the producer. When the milk buyer has made an error while filling in the quantity supplied he should not be allowed to alter the figure but he should cancel the receipt and write a new one. The second copy of a cancelled receipt should remain in the book.

Whatever system is used for recording milk purchases one should always avoid that the milkbuyer has to record the quantity supplied twice; once on a record kept by the producer and once on an accounting record. If the milk buyer writes a different figure on each of the documents this will create problems when the producer is paid and he could lose confidence in the whole dairy organisation.

Every time the milk buyer has completed one page of milk receipts, he should add the quantities supplied recorded on that page and fill in the amount in the space "total of this page" the cumulative total of the previous page should be entered in the space "brought forward" and the new cumulative total should be calculated. The cumulation of totals should start from the first receipt made after the lorry transporting milk to the processing plant, has left the collection centre. For cooling centres collecting evening milk, the accumulated quantity should include the evening milk of one day and the morning milk of the next day.

When the purchase operations are closed the milk buyer should cancel the unused milk receipts of the last page used during the day and he should calculate the grand total of milk purchased ready for dispatch to the processing plant. Each milk can to be dispatched to the processing plant should be marked with the collection centre's code.

Accurate handing-over of milk from the milk buyer to the driver from the processing plant would be time consuming. The value of the milk lost during the process could be considerably higher than the benefit obtained from a very accurate handing-over, especially when there are several collection centres without cooling facilities along the same route.

If the milk buyer has filled the cans to capacity, only the number of full cans need to be counted and multiplied with their capacity. The volume of milk in the one can that is not full should be estimated or measured with a dipstick, if available. This should be recorded on a Delivery Note made in duplicate. The milk buyer should record the various capacities of the milk cans in the column "litres per can". Since the capacity of milk cans is usually given in litres, it is easier if this part of the delivery note is completed in litres.

The work of the milk buyer would also be simplified if a standard type of cans could be used. The one can that is not completely full should be recorded as a can with a different capacity than the others. The milk buyer should further enter the number of cans of each capacity and calculate the total litres by multiplying the litres per can with the number of cans. He should then add the figures in the column "total litres" to obtain the total number of litres handed over to the driver.

The driver should sign the delivery note after checking that the number and types of cans loaded on the lorry agree with the entries on the delivery note. The first copy of the delivery note and the first copy of all used milk receipts should be handed over to the driver.

Reception at the Processing Plant

On arrival at the processing plant the driver should give the milk receipts and delivery notes to the receptionist. The milk cans should be off loaded on to the platform or a chain conveyor, taking care that the cans from each collection centre remain together. The milk which is accepted is tipped in the weighbowl and the quantity in kilograms should be recorded by the receptionist on the delivery note of the collection centre from where the milk originated. The volume of the weighbowl is usually larger than the content of one can; more cans may be tipped in the weighbowl before the quantity need to be recorded, as long as the milk from different collection centres is weighed separately. The receptionist should also count the number of cans received from each collection centre and this number should agree with the number of cans dispatched by the collection centre as recorded on the delivery note.

The quantity of rejected milk should be recorded in the appropriate space of the delivery note. The receptionist should then calculate the total fresh milk and the total milk received at the processing dairy. He should also convert the "total litres received by the driver" to kilograms. The total milk received at the dairy (including rejects) should be compared with the total kilograms received by the driver at the collection centre.

The delivery note only allows to check on the drivers, but it is even more important to know if there is any difference between the quantity of milk purchased at each collection centre, which will be paid for, and the quantity of milk received at the processing dairy, which will eventually become available for sale.

This should be controlled by using the "Reception Quantity Control Sheet" to be maintained in duplicate by the receptionist. When the number of routes and collection centres is not too large only one sheet would be sufficient. In this case the form should be subdivided to allow space for totals of each collection route. When the number of collection centres would result in too large forms one reception quantity control sheet should be kept for each collection route. Whether one or more forms need to be used each day the procedure remains the same.

Before completing the reception quantity control sheet the receptionist should check the calculations made by the milk buyers on the milk receipts. He should then transfer the quantity purchased at each collection centre from the milk receipts to the reception quantity control sheet. The figures for the columns: "delivered to driver", "total received at dairy" and "reject" should be obtained from the delivery notes. The delivery note number should be entered in the appropriate column to facilitate any verification required later on.

The receptionist should then calculate the difference between the milk purchased and the milk received at the dairy. Because the figures used to calculate this difference are both obtained by weighing it gives the most accurate base for evaluating the efficiency of the purchase and transport operations. When the difference between the quantities purchased and received at the dairy for any one collection centre is important, all calculations on the milk receipts and delivery note for that collection centre should be checked before any conclusions are drawn. If after checking the difference is still significant it should normally be a shortage. If a surplus is found it would indicate that milk was bought at the collection centre but that no receipt was issued for it.

The difference between the milk purchased and the milk delivered to the driver need to be calculated only when the difference between milk purchased and milk delivered to the dairy is of any significance. In case such an important shortage occurs the reception quantity control sheet allows to identify who should be held responsible: the milk buyer at the collection centre or the driver.

The receptionist should further calculate the quantity of fresh milk available for further processing. This is obtained by deducting the quantity of rejected milk from the quantity of milk received at the dairy.

When the receptionist has entered the figures for all collection centres on the reception quantity control sheet he should calculate for each column sub totals per route and the grand total for the day. The difference between the milk purchased and the milk received at the dairy and also the daily total quantity of rejected milk should be calculated as a percentage of daily total quantity of milk purchased.

Because the measuring devices on the storage tanks in the processing sections usually show litres and because sales of most products are made in litres it is necessary that the receptionist converts the daily total of fresh milk from kilograms to litres. He should also calculate the total fat in the fresh milk using the average fat percentage obtained by the laboratory.

The daily totals should then be transferred to the reception control summary sheet. This document should be kept by the receptionist and covers the milk purchase and reception for a whole month. This document gives a better idea in the change taking place from day to day in the purchase and reception operations than could be obtained by looking through a file with the daily reception quantity control sheets.

On completion of these documents the receptionist should forward the milk receipts and one copy of the reception quantity control sheet to the accounts office where the accounts for the individual producers are kept. After copying the quantities recorded on the milk receipts to the appropriate producer accounts a total of all entries on the producer accounts should be made and this total should agree with the total milk purchased as recorded on the reception quantity control sheet.

At the end of the month the total quantity purchased as recorded on the producer accounts should agree with the monthly total of milk purchased as recorded on the reception control summary sheet. No payment calculation should be made before both totals agree.

Pasteurisation Section

In the dairy plant under consideration the pasteurisation section is responsible for the following activities:

- — storage of raw milk
- — reconstitution and/or recombination
- — pasteurisation
- — homogenisation
- — separation
- — distribution of milk to the processing units

The pasteurisation section receives in the first instance raw milk from the reception. Since the milk flows through a closed system from the dumptank in the reception to the storage tanks in the pasteurisation section the quantity weighed in at the reception should normally agree with the quantity of raw milk received by the pasteurisation section.

'The total quantity of milk received in litres as calculated on the reception quantity control sheet should be entered on the pasteurisation control sheet. The readings on the storage tanks should only be used as a control. Only when a flow meter is available should the quantity measured by this device be recorded on the pasteurisation control sheet. The figures for fat content and total fat should be obtained from the reception quantity control sheet.

In many countries the raw milk supply does not meet the demand for liquid milk, at least at certain times of the year. Where this is the case, liquid milk has to be produced from milk powder and butteroil. The pasteurisation section being responsible for reconstitution, irregular mixing of the milk powder should be shown in the final result of this section.

The total volume and the quantity of fat in the butteroil used for recombination should be recorded on the "in" side of the pasteurisation control sheet. The opening stock to be recorded is the closing stock of the previous day.

After pasteurisation the milk is immediately distributed to the reception tanks of the processing sections and the quantity "out" should be read on the measuring device of these tanks. The fat content of the whole milk on the "out" side of the pasteurisation control sheet should be the same as that for the raw milk recorded on the "in" side. The fat content for cream and for recombined milk should be obtained from the laboratory and the total quantity of fat should be calculated.

The closing stock of the pasteurisation section should only consist of unpasteurised products since all pasteurised milk is stored in the processing sections. At the end of each day the columns "litres" and "fat kg" should be totalled and the difference between "in" and "out" should be calculated for the total volume and the total quantity of fat.

Liquid Milk Section

The two main functions of the liquid milk section are the standardisation of the milk so as to obtain the fat content required in the final product and packaging that product in containers. The dairy plant produces two types of liquid milk; standard milk with 3% fat and low fat milk with 1.5% fat. Part of the milk is packed in single service containers of 1L and ½L and these packages are delivered in returnable crates. Another part of the milk is packaged in returnable 45L cans for distribution to hospitals, hotels etc.

The opening stock to be entered on the "in" side of the liquid milk control sheet are the quantities of pasteurised milk left over-night in the storage tanks of the liquid milk section. The quantities to be entered as "received from pasteurisation section" should be read from the measuring device on the storage tanks of the section. These quantities as well as the fat content should be the same as the entries "to liquid milk section" on the "out" side of the pasteurisation control sheet.

On the "out" side of the liquid milk control sheet the number of units of each size and type should be recorded and converted into litres by multiplying the number of units with the corresponding unit size. The units should be counted at the time they enter the coldroom when they come under the responsibility of the storekeeper.

At the end of each day the difference between total "in" and total "out" should be calculated and the loss or surplus should be calculated as a percentage of the total "in". This last figure for the number of litres as well as the quantity of fat indicate the efficiency of the liquid milk section.

Fermented Milk Section

The fermented milk section receives various types of milk from the pasteurisation section to be blended to the correct composition for producing the type of fermented milk required. The fermented milk section may also receive buttermilk from the butter section.

If any stock is carried over-night by the fermented milk section this should be entered as opening stock on the fermented milk control sheet. On the "in" side of the control sheet should further be recorded: the volume and fat content of the various types of milk received from other sections, starter and the volume of non-milk additives such as sugar, fruitpulp etc.

The figures to be recorded on the "out" side of the fermented milk control sheet should be obtained in the same way as for the liquid milk section as well as the difference and the loss/surplus percentage.

Butter Section

In the processing sections studied so far the total volume and the total fat used should also be found in the final product. This is not the case in the manufacture of butter, cheese, ghee etc. In these processes only part of the input is found in the final product and byproducts are produced, some of which can be marketed.

The documents to be used for the quantity control of these processes would be similar, they would have three sections: inputs, manufacturing and packaging. As far as quantity control is concerned three stages in the butter manufacturing are important: cream ripening (inputs), churning (manufacturing) and packaging.

In dairy plants where butter making is a daily operation there should always be ripened cream, prepared the previous day opening stock. During the day the butter section receives fresh cream from the pasteurisation section to which starter is added. The addition of these inputs gives the total available. The ripened cream transferred to the butter churn should be deducted from the total above to obtain the closing stock.

Each time the churn is filled with ripened cream the quantity should be recorded in the churning section of the butter control sheet. The quantity of butter and butter milk obtained should also be recorded for each churn. The daily totals should be calculated for: cream used, butter and butter milk and the yield can then be calculated. The butter milk is issued to the liquid or the fermented milk section depending on the way it will be used. The quantity of butter milk issued by the butter section should also be found on the "in" side of the control sheet of the receiving section, in this example the fermented milk section.

The total quantity of butter produced should be entered as received on the packaging side. This quantity added to the stock of bulk butter not packaged the previous day gives the total available for packaging. The number of units packaged should be counted when entering the coldroom and should be multiplied by the unit weight to obtain the total kg packaged. To this should be added the closing stock of bulk butter and this should be compared with the total available. When the bulk butter kept over-night does not fall under the responsibility of the butter section, the headings opening and closing stock on the butter control sheet should be replaced by "received from coldroom" and "issued to coldroom" respectively. The coldroom control sheet should then also have space for recording the reception and issues of bulk butter.

Sales Section

The system for sales control described below is generally known as the pegboard system. It requires a board with a line of pegs at the top. Most boards available in commerce are also equipped with a sliding ruler and a device to put the board at different inclinations to facilitate the reading of the documents placed on the board. The documents to be used with the system need to have holes punched at the top so that they can fit in the pegs of the board. Because many calculations need to be made horizontally when using the peg board system the use of an adding machine or a calculator is necessary.

All products sold by the dairy plant should be printed on the documents, one line should be used for each unit size of each product. The documents should be printed in such a way that when the documents are fitted on the peg board the line for one particular product and size is at the same level on all the different types of documents. When putting these documents on the peg board each document should cover part of the previous one and only the column with the figures required should remain uncovered. When the figures that remained uncovered are added horizontally they give the total product by product.

The main advantage of the peg board system is that the figures of the original document need not to be copied on a summary sheet; a source of errors is thus eliminated and time saved. An additional advantage of the system is that, since all products are printed on invoices and delivery notes, the customers are constantly reminded of the full range of products offered for sale by the dairy plant.

The dairy plant for which the system is worked out in more detail employs its own sales-men who sell to retail shops not owned by the dairy organisation.

The sales manager should prepare one store issue voucher in triplicate for each salesman. The last copy of the store issue voucher should remain in the book, the two other copies should be forwarded to the storekeeper who should prepare the consignments for each salesman as indicated on the issue voucher. The storekeeper should record the number of returnable crates and cans in the appropriate space of the issue voucher.

When the goods are handed over to the salesman he should sign the two copies of the issue voucher, one of these copies should remain with the storekeeper. The salesman should keep the other copy for control at the gate of the dairy plant. No products should leave the plant without a store issue voucher. The sales manager may delegate the responsibility for completing the store issue voucher to an other staff member but this responsibility should never be given to a storekeeper because the store issue voucher is an authority for the storekeeper to issue the products recorded on it.

When the goods are delivered to the customer the salesman should prepare a delivery note if the customer is allowed to buy on credit or a cash sale. These documents should be made in triplicate. The customer should sign all three copies of the delivery note while the salesman should sign all copies of the cash sale when he has received the payment. The first copy of the delivery note or cash sale should be given to the customer, the other copies should remain in the book.

When the salesman has completed his route he should bring the unsold products to the coldroom and give his copy of the store issue voucher to the storekeeper. The storekeeper should record the returned products and containers on the two copies of the store issue voucher and he should calculate the net sales of each product.

One copy of the store issue voucher should remain with the storekeeper and the salesman should take the other copy together with the delivery notes and cash sales to the sales manager. The sales manager should first check if no alterations were made on the salesman's copy of the issue voucher by comparing it with the copy that remained in the book.

The sales manager should then take the second copy of the delivery notes and cash sales out the book and put them on the peg board with the quantity column uncovered. He should add the quantities recorded on the delivery notes and cash sales for each product (line by line) and these totals should agree with the net quantity for each product recorded on the store issue voucher. The delivery notes and cash sales should then be replaced on the peg board so that the columns with the amounts remain free. The amounts should be added line by line and the results recorded on the store issue voucher.

The following checks should then be carried out:

- The net quantity of each product as recorded on the store issue voucher multiplied by the unit price should equal the amount obtained by the addition of delivery notes and cash sales.
- The addition (vertical) of the amounts per product on the store issue voucher should agree with the addition (horizontal) of the total amounts of all delivery notes and cash sales.

The net quantity of containers can be a negative figure when more containers were returned than delivered. This should be clearly indicated, e.g. by making a ring around negative figures. The sales manager should further calculate the total amount of credit sales and the total amount of cash sales and record them in the appropriate space of the store issue voucher.

The salesman should pay the cash collected during his tour to the cashier who should issue a receipt. The salesman should bring his copy of the cash receipt to the sales manager who would record the amount of the receipt on the store issue voucher. If this amount agrees with the total of the cash sales as calculated previously the process is completed. When there is a difference between the two amounts action should be taken depending on the policy of the dairy organisation in such matters. On the store issue voucher space has been provided for showing an eventual shortage or overage of the salesman. This is necessary in cases where shortages are deducted from the salesman's pay. Surplusses could be used for offsetting shortages made by the salesman or they could be considered as miscellaneous revenue for the dairy plant.

When all salesmen have returned to the plant and their store issue vouchers are checked the sales manager should put all store issue vouchers on the peg board and calculate the daily sales summary. These daily sales summaries can later be used for preparing weekly and/or monthly summaries.

The store issue vouchers should be filed per salesman. At the end of a week or a month each salesman's vouchers should be put on the peg board to calculate his total sales, shortages or surplusses. The monthly total sales can be used for comparing each salesman's performance with that of his colleagues or with the performances of previous periods. It can also be used as a basis for calculating each salesman's sales commission where this is applicable.

When besides the wholesale operations described above, the dairy organisation also operates its own retail outlets the procedures would be similar. The main difference would be that no delivery notes or cash sales are used and the control of this operation should be done with the store issue vouchers only. Daily, weekly and monthly sales summaries should be made separately for wholesale and retail operations otherwise the control based on the quantity-price relation would be lost.

For dairy plants that bring a large range of products on the market the documents used with the peg board system may become too long. In this case each sales document should be split into two similar documents each with different products. Products with a long shelf life would in most cases be distributed less frequently than products with a short shelf life and this could be used as a criterion for separating the products.

Coldstore Stock Control

The coldroom receives products from two sides: the production sections of the plant and unsold returns from the salesmen. The entries into the coldroom from the production sections should be the same as the quantities "out" recorded on the control sheet of each production section.

The quantities returned by the salesmen can be obtained by adding the return column of the store issue vouchers of all salesmen. The storekeeper should put the copy of the store issue vouchers that remained in the coldroom on the peg board to make this addition. The returns should be recorded separately from the fresh products on the coldroom control sheet. This is necessary because it is usually not the storekeeper's responsibility to decide what should be done with the returns, but as long as they remain in the coldroom he is responsible and they need to be recorded.

When a decision has been made as to how the returns will be used, normally the next day, the quantities that can still be used should be recorded as "other issues" on the coldroom control sheet and as received in the section where they will be used. If the returns would be used by more than one production section this should be specified on the coldroom control sheet and more lines should be provided for "other issues". Returns that cannot be used should be recorded as leakages.

If the keeping quality of the product allows the returns to be sold as fresh products the quantity of these returns should first be recorded as "other issues" in the return column and entered as received in the fresh product columns of the coldroom control sheet.

The sales issues of fresh products should be obtained by adding the "quantity issued" column of all store issue vouchers issued during the day. "Other issues" of fresh products may be products taken for laboratory tests or products distributed free of charge to visitors or staff. When all figures are filled in on the coldroom control sheet, the storekeeper can calculate the total intake, issues and closing stock in units and in litres.

Regularly a senior staff member should do a physical stock-taking and record his findings on the line "physical stock". If there is any difference between the physical stock and the calculated stock this should be recorded on the coldroom control sheet and appropriate action should be taken. When stock-taking has been done the physical stock should be carried forward as opening stock for the next day and not the calculated stock, otherwise the difference would accumulate.

REFERENCES

Carawan, R. E., *Processing Plant Waste Management Guidelines—Aquatic Fishery Products,* North Carolina Pollution Prevention Program, 1991.

Harper, W. F. and Blaisdell, J.L., "State of the art of dairy food plant wastes and waste treatment", *Proceedings of the Second National Symposium on Food Processing Wastes,* Denver (Colorado), USA, 23–26 March, 1971.

Robinson, R. K., *Modern Dairy Technology, Volume 1.* Elsevier Applied Science Publishers, London, 1986.

Varnam, A. H. and Sutherland, J. P., *Milk and Milk Products: Technology, Chemistry and Microbiology,* Chapman and Hall. London, 1994.

3

Dairy Development

In more than 40 years of independence, experiments have been made with a number of approaches towards developing the dairy industry. In India, dairying is recognised as an instrument for social and economic development. The nation's milk supply comes from millions of small producers, dispersed throughout the rural areas. These farmers maintain an average herd of one or two milch animals, comprising cows and/or buffaloes. The animals' nutritional requirements are largely met by agricultural waste and by-products. Ample labour and a small land base encourage farmers to practice dairying as an occupation subsidiary to agriculture. While income from crop production is seasonal, dairying provides a stable, year-round income, which is an important economic incentive for the small farmer to take to dairying.

Milk production in India is dominated by small and marginal landholding farmers and by landless labourers who, in aggregate, own about 70 percent of the national milch animal herd. As crop production on 78 percent of the agricultural land still depends on rain, it is prone to both drought and floods, rendering agricultural income uncertain for most farmers. Shackled to subsistence production as a result of a shortage of finance and credit facilities, these farmers become entangled in a strangling debt cycle.

The combination of an unfavourable land: person ratio and fragmented landholdings makes it difficult to support large families on crop income alone. It is not surprising that the small farmer and the landless labourer are more often than not victims of moneylenders and of natural calamities. Dairying, as a subsidiary source of income, is a real relief to most of these weaker groups in society. Often one or two milch animals enable these farmers to generate sufficient additional income to break the vicious subsistence agricultural-debt cycle.

The successful Indian dairy development programme Operation Flood has shown how food aid can be used as an investment in building the type of institutional infrastructure that can bring about national dairy development. Programmes like Operation Flood, with similar policy orientations, may prove to be appropriate to dairy development in other Asian as well as African countries since the conditions that prevail in dairying today in a number of developing countries are comparable to those that once were found in India.

In the early 1950s, India was commercially importing around 55000 tonnes of milk powder annually to meet the urban milk demand. Currently, many developing countries either commercially import dairy products on a large scale or rely on donations to meet the gap between demand and supply of milk products. Together, developing-country imports account for over 70 percent of total world trade in milk products.

As in India, the reliance on low-cost and often-subsidised commercial imports as well as gifts seriously affects the development of dairying systems in many developing countries as they increase domestic demand for milk products and erode domestic-price incentives. As low-cost imports from advanced dairying nations depress domestic milk prices, it becomes uneconomical for local milk producers to invest in dairying, causing stagnation in production. This hampers breed improvement efforts, fodder production programmes, the manufacture of quality feed and other endeavours to improve productivity. For any country seriously interested in exploiting the potential of dairying, therefore, it becomes imperative to review and seriously consider policies that would promote dairy development activities by helping to establish independent and self-sustaining dairying systems.

Undoubtedly, the major challenge for the dairy sector in any developing nation is to increase milk production in order to meet the increasing demand resulting from the almost inevitable expansion of population and, presumably, growth of income. To meet this challenge, policies must become more market-oriented. The adoption of appropriate technologies for production, procurement, processing and marketing—after the unique environmental, social, economic, political and cultural environment of the individual country has been considered—is an important aspect of dairy development. Those national governments and international institutions for whom the dairy sector is a major concern should accept the challenge and formulate policies that integrate and buttress the major functions of dairy development.

Most of the significant developments in dairying in India have taken root in this century. The history of the dairy development can be broadly classified into two distinct phases: pre- and post-Operation Flood. On examining the developments made during both phases, it can be concluded that the difference lies in the basic approach to solving the problems facing the dairy sector.

The earliest attempts at dairy development can be traced back to British rule, when the Defence Department established military dairy farms to ensure the supply of milk and butter to the colonial army. The first of these farms was set up in Allahabad in 1913; subsequent facilities were established at Bangalore, Ootacamund and Karnal. These farms were well maintained and, even in the early stages, improved milch animals were raised. As animals were reared under farm conditions, some herd improvement was made using artificial insemination. This approach did not have any impact on the supply of milk to urban consumers, which was of major concern to civilian authorities but less important to the military.

With the growth of the population in urban areas, consumers had to depend on milk vendors who kept cattle in these areas and sold their milk, often door-to-door. As a result, several cattle sheds came into existence in different cities. This was not an environmentally sound approach. As the main objective of the milk vendors was to maximise profit, they started increasing the lactation period by using the focus system. In the process, these high-yielding cattle developed sterility problems, which considerably reduced the number of carvings. Once the cattle became unproductive, they were sold to slaughterhouses. This practice systematically drained the country of its genetically superior breeds.

To some extent, the Second World War gave impetus to private dairies with modestly modernised processing facilities. In the cities of Bombay, Calcutta, Madras and Delhi, and even in some large townships, processed milk, table butter and ice-cream were available, though not on a large scale. Polsons, Keventers and the Express Dairy were some of the pioneer urban processing dairies. These dairies were not concerned with improving the breed of milch animals reared in rural pockets but instead were content with contracting milk supplies through middlemen or their own staff. Milk producers as well as consumers were exploited. These early modern systems did not bring about significant shifts in milk production, nor did they develop quality milch animals. To a large extent, despite modernised processing facilities, dairying remained unorganised.

With the initiation of India's first Five-Year Plan in 1951, modernisation of the dairy industry became a priority for the government. The goal was to provide hygienic milk to the country's growing urban population. Initial government action in this regard consisted of organising "milk schemes" in large cities. To stimulate milk production, the government implemented the Integrated Cattle Development Project (ICDP) and the Key Village Scheme (KVS), among other similar programmes. In the absence of a stable and remunerative market for milk producers, however, milk production remained more or less stagnant. During the two decades between 1951 and 1970, the growth rate in milk production was barely 1 percent per annum, while per caput milk consumption declined by an equivalent amount.

During the 1960s, various state governments tried out different strategies to develop dairying, including establishing dairies run by their own departments, setting up cattle colonies in urban areas and organising milk schemes. Almost invariably, dairy processing plants were built in cities rather than in the milksheds where milk was produced. This urban orientation to milk production led to the establishment of cattle colonies in Bombay, Calcutta and Madras. These government projects had extreme difficulties in organising rural milk procurement and running milk schemes economically, yet none concentrated on creating an organised system for procurement of milk, which was left to contractors and middlemen.

Milk's perishable nature and relative scarcity gave the milk vendors leverage, which they used to considerable advantage. This left government-run dairy plants to use large quantities of relatively cheap, commercially imported milk powder. The daily per caput availability of milk dropped to a mere 107 g during this time. High-fat buffalo milk was extended with imported milk powder to bring down the milk price, which resulted in a decline in domestic milk production. As the government dairies were meeting barely one-third of the urban demand, the queues of consumers became longer while the rural milk producer was left in the clutches of the trader and the moneylender.

All these factors combined left Indian dairying in a most unsatisfactory low-level equilibrium. The establishment and prevalence of cattle colonies emerged as a curse for dairying in the rural hinterland as it resulted in a major genetic drain on the rural milch animal population, which would never be replaced. City dairy colonies also contributed to environmental degradation, while the rural producer saw little reason to increase production.

Dairy Development in Rural and Urban Areas: Amul and the evolution of the Anand model

Milk procurement from the rural areas and its marketing in the urban areas was the major problem in Indian dairying at the time India gained independence. In one of the earliest urban milk supply schemes, Polsons—a private dairy at Anand—procured milk from milk producers through middlemen, processed it and then sent the milk to Bombay, some 425 km away. Bombay was a good market for milk and Polsons profited immensely. In the mid-1940s, when the milk producers in Kaira asked for a proportionate share of the trade margins, they were denied even a modest increase. The milk producers went on strike, refusing to supply milk to Polsons. On the advice of Sardar Vallabhbhai Patel, a leader in India's independence movement, the milk producers registered the Kaira District Cooperative Milk Producers' Union, now popularly known as AMUL, in 1946. The Kaira union procured milk from affiliated village-level milk societies. This was the genesis of organised milk marketing in India, a pioneering effort that opened a new vista for dairy development in the country.

Between 1946 and 1952, AMUL's policy was directed towards obtaining monopoly rights for the sale of milk to the Bombay milk scheme. In 1952, it succeeded in achieving its purpose after the Government of Bombay cancelled the contract with Polsons and handed over the entire business of supplying milk from the Kaira district to AMUL. However, as the Bombay milk scheme was committed to purchasing all the milk produced by the Aarey Milk Colony in Bombay, it would not take AMUL's milk during the peak winter months. The disposal of this surplus milk posed difficulties for AMUL, forcing it to cut down on purchases from its member societies, which affected members' confidence. The answer was the production of milk products: in 1955, a new dairy plant was set up at Anand to produce butter, ghee and milk powder.

A second dairy was built in 1965, and a product manufacturing unit was established in 1971 to cope with increasing milk procurement. In 1993, a fully automatic modern dairy was constructed adjacent to the original AMUL dairy plant at Anand.

AMUL formed the basis for the Anand Model of dairying. The basic unit in this model is the milk producers' cooperative society at the village level. These cooperatives are organisations of milk producers who wish to market their milk collectively. Membership is open to all who need the cooperative's services and who are willing to accept the responsibilities of being a member. Decisions are taken on the basis of one member exercising one vote. No privilege accrues to capital, and the economic returns, whether profit or loss, are divided among the members in proportion to patronage. Each cooperative is expected to carry out the continuing education of its members, elected leaders and employees. All the milk cooperatives in a district form a union that, ideally, has its own processing facilities.

All the unions in a state are normally members of a federation whose prime responsibility is the marketing of milk and milk products outside the state. There is also a fourth tier, the National Cooperative Dairy Federation of India (NCDFI), which is a national-level body that formulates policies and programmes designed to safeguard the interests of all milk producers. Each tier of the Anand organisational structure performs a unique function: procurement and services by the cooperative; processing by the union; marketing by the state federation; and advancing the interests of the cooperative dairy industry by the national federation. Thus, the Anand Model has evolved into an integrated approach to systematic dairy development.

Operation Flood Programme

The strategy for organised dairy development in India was actually conceived in the late 1960s, within a few years after the National Dairy Development Board (NDDB) was founded in 1965. It rested on the Operation Flood programme, which was conceived by the NDDB and endorsed by the government.

Operation Flood is a unique approach to dairy development. During the 1970s, dairy commodity surpluses were building up in Europe, and Dr Verghese Kurien, the founding chairman of NDDB, saw in those surpluses both a threat and an opportunity. The threat was massive exports of low-cost dairy products to India, which, had it occurred, would have tolled the death-knell for India's staggering dairy industry. The large quantities that India was already importing had eroded domestic markets to the point where dairying was not viable.

The opportunity, on the other hand, was built into the Operation Flood strategy. Designed basically as a marketing project, Operation Flood recognised the potential of the European surpluses as an investment in the modernisation of India's dairy industry. With the assistance of the World Food Programme, food aid—in the form of milk powder and butter oil—was obtained from the countries of the European Economic Community (EEC) to finance the programme. It was the first time in the history of economic development that food aid was seen as an important investment resource. Use of food aid in this way is anti-inflationary, it provides a buffer stock to stabilise market fluctuations and it can be used to prime the pump of markets that will later be supplied by domestic production. The overriding objective of all aid is, or rather should be, the elimination of the need for aid. The use of food aid as an investment is the most effective way of achieving this objective.

Operation Flood is a programme designed to develop dairying by replicating the Anand Model for dairy development, which has stood the test of time for almost half a century. The first phase of Operation Flood was launched in 1970 following an agreement with the World Food Programme, which undertook to provide as aid 126000 tonnes of skim milk powder and 42000 tonnes of butter oil to finance the programme.

The programme involved organising dairy cooperatives at the village level; creating the physical and institutional infrastructure for milk procurement, processing, marketing and production enhancement services at the union level; and establishing dairies in India's major metropolitan centres. The main thrust was to set up dairy cooperatives in India's best milksheds, linking them with the four main cities of Bombay, Calcutta, Delhi and Madras, in which a commanding share of the milk market was to be captured. In achieving that goal, the first phase of Operation Flood laid the foundation for India's modern dairy industry, an industry that would ultimately meet the country's need for milk and milk products.

The second phase of the programme was implemented between 1981 and 1985. Designed to build on the foundation laid in the first phase, it integrated the Indian Dairy Association-assisted dairy development projects being implemented in some Indian states into the overall programme. About US$150 million was provided by the World Bank, with the balance of project financing obtained in the form of commodity assistance from the EEC.

The current, third phase of Operation Flood aims at ensuring that the cooperative institutions become self-sustaining. With an investment of US$360 million from the World Bank, commodity and cash assistance from the EEC and NDDB's own internal resources, the programme envisages substantial expansion of the dairy processing and marketing facilities; an extended milk procurement infrastructure; increased outreach of production enhancement activities; and professionalisation of management in the dairy institutions.

National Dairy Development Board

In October 1964, on the occasion of the inauguration of AMUL's cattle feed plant, the then Prime Minister of India, Lal Bahadur Shastri, spent the night as the guest of a village milk cooperative society near Anand. Impressed by the socio-economic changes brought about by the milk cooperatives, he expressed the desire for a national-level organisation to replicate Anand Model dairy cooperatives throughout the country and to make available multidisciplinary, professional dairy expertise to dairies in the public and cooperative sectors. Thus, in 1965, NDDB was registered under the Societies Registration Act, the Charitable Trust Act and the Public Trust Act. Consistent with its model and mandate, NDDB headquarters were established at Anand.

During its initial stages, NDDB was assisted financially by the Government of India, the Danish Government and by AMUL. It also received aid from the United Nations Children's Fund (UNICEF) in the form of teaching material and equipment.

In 1969, when the Government of India approved the Operation Flood programme and its financing through the monetization of World Food Programme-gifted commodities, it was found that the statutes under which NDDB was registered did not provide for handling of government funds. Therefore, in 1970 the government established a public-sector company, the Indian Dairy Corporation. The IDC was given responsibility for receiving the project's donated commodities; testing their quality; their storage and transfer to user dairies; and receiving the dairies' payments. Thus, it served as a finance-cum-promotion entity while the entire Operation Flood technical support was provided by NDDB.

To avoid-any duplication in their activities or overlap of functions, the IDC and NDDB were eventually merged into a newly constituted NDDB by an Act of Parliament passed in October 1987. The Act designated the NDDB as an institution of national importance and accorded it the same autonomy of operation that it had enjoyed and which had been a major factor in its success.

Operation Flood may be considered the central event of twentieth-century dairying in India. An analysis of the lessons learned through the implementation of the programme should be useful for those involved in formulating dairy development policies and programmes for the developing nations of Asia and Africa.

The network of cooperative institutions created through the Operation Flood programme now comprises 70000 dairy cooperative societies in 170 milksheds, encompassing 8.4 million milk-producer families. Average milk procurement by these cooperatives has now reached some 12.3 million kg per day, of which 8.2 million litres are marketed as liquid milk, while the remainder is converted into products such as milk powder, butter, cheese, ghee and a wide range of traditional milk products. Milk-processing capacity of approximately 15.6 million litres per day, chilling capacity of 6.5 million litres per day and milk powder production capacity of 726 tonnes per day have been established through the programme.

One of the challenging aspects of dairy development in a tropical or subtropical country is the movement of milk over long distances. In Operation Flood, this has been made possible through the operation of about 140 insulated rail milk tankers, each with a capacity of 40000 litres, supplemented by another 25 rail tankers of 21000-litre capacity. Approximately 1000 other insulated road milk tankers operate throughout the country as well. This has enabled the operation of a national milk grid, balancing regional fluctuations in milk procurement and demand-and-supply gaps resulting from concentrated production of liquid milk in selected milksheds. To balance seasonal variations in milk supply and demand caused by low milk production during the summer months, a large milk powder storage capacity has been created for buffer stocking.

The investment and achievements in modernising the Indian dairy industry have had a major impact on milk production. Annual production, which had stagnated to between 20 million and 22 million tonnes during the 1960s, has steadily increased to around 59 million tonnes, an annual growth rate of about 7.8 percent. Per caput availability of milk, which had declined consistently during the two decades between 1951 and 1970, dropping to 107 g at the start of Operation Flood, is now 187 g per day, despite a substantial increase in population. Had family-planning programmes achieved the same success as Operation Flood, the per caput consumption rate would be comparable to all but those of the leading dairying nations.

Commercial imports of dairy commodities were a regular feature in the 1950s and 1960s, comprising 50 to 60 percent of the dairy industry's total throughput. Today, imports of dairy commodities are restricted to those donated by the EEC for implementing Operation Flood and their percentage of the total dairy throughput is negligible. All these developments have helped raise India to the rank of the second-largest producer of milk in the world, next only to the United States.

A number of programmer and policies have played a role in this success. Certainly, the introduction of modern technology, both at the farmer level and in the processing of milk and products, has been important. Similarly, establishing an urban market has provided the stability necessary to encourage farmers to invest in increased milk production. The induction of professional managers to serve farmers has reversed the usual pattern of farmers as supplicants and officials as "benefactors". Perhaps most important, however, is the cooperative structure itself. By giving farmers command over the resources they create, Operation Flood has ensured that they receive the maximum return from each rupee spent by consumers on milk and milk products, and it is this that has provided the incentive on which the growth of the dairy industry has been based.

The success of Operation Flood has resolved many difficult issues relating to development. It has demonstrated how food aid can be used to enhance domestic production if administered with care. It has also shown how technology can be harnessed so that neither the dependence on imported technology nor its capital intensity become counterproductive. Some of the dairy plants set up by NDDB during the implementation of Operation Flood are based on the latest technology and are comparable to those in advanced countries. The unique cooperative infrastructure with which NDDB works makes the adoption of technologies and the dissemination of knowledge relatively easy, and this has enabled Operation Flood to facilitate the application of modern technologies to enhance milk production.

After a long struggle, India has overcome a situation that, at least in some respects, may be similar to those prevailing in a number of Asian and African countries and has built a modern dairy sector, responsive to the needs of milk producers. It has had to deal with a stagnant dairy industry in which cheap, subsidised imports were a disincentive to the farmers. It has had to overcome the negative effects of consumer-oriented programmer that managed to keep prices low for the urban elite while depressing the price of milk in rural milksheds. It has had to deal with the lethargy and bureaucratic orientation of state enterprise in dairying.

While there are still challenges to be met, the foundation for an even stronger dairy industry is now in place. Early in the last decade, NDDB was asked to explore the possibilities of introducing the Anand Model into a number of countries and has responded by sending investigative teams into those countries. FAO, EEC and the World Bank, having recognised the potential of the Anand Model, are considering financing similar projects in Sri Lanka, Malaysia and Bangladesh. Commodity financing of projects such as Operation Flood by the more advanced countries not only relieves some of their accumulated surplus, but it also strengthens the economies of the countries that receive assistance. And, if we truly believe that the proper objective of aid is to end the need for aid, then this is certainly an important way of achieving that goal.

Cooperative Structure in Dairy Development

As a result of the perishable nature of milk and the range of skills involved in its production and marketing, dairying requires a number of services that can best be provided by cooperative action. It is not surprising therefore that the cooperative movement has featured prominently in the development of the dairy industry worldwide.

The aim of a producers' cooperative is to provide services either free of charge or at a reasonable cost to its members.

In addition, cooperative ownership emphasises participation and control by member producers Individual members can influence policy and management matters through registered membership bodies that are regulated by laws or rules of a community or state.

The cooperative principles as enunciated by the International Cooperative Alliance are voluntary and open membership; democratic control; limited interest on credit; equitable division of surplus; training of members; and cooperation among cooperatives.

Cooperative organisations are aware of the importance of member involvement and, through training programmes, try to increase participation in the affairs of the society. The fact that each member has only one vote is particularly important; in the case of public companies, individual shareholders find it difficult to effectively control the management of a company unless they have a controlling interest through ownership of a large number of shares.

As is generally well known, land is perhaps the most important income-generating asset in the rural economies of Asia. Yet, scarcity of land and its skewed distribution are two of the major constraints of the rural Asian landscape. Close to 60 percent of the world's agricultural population lives in the villages of Asia, but its share of the world's agricultural land is only about 28 percent. This is much lower than in other developing regions.

Furthermore, of this limited available land, a very large portion is comprised of holdings other than small farmers' holdings. While the conditions vary from country to country, small farmers in Asia, although accounting for more than two-thirds of the rural households, have access to only about 20 percent of the arable land. Against this background, economic activities that are not essentially land-based, such as dairying, have become crucial for small farmers and landless labourers.

While farmers' cooperatives of various types play a useful role in promoting rural development, dairy cooperatives have special attributes that make them particularly suitable.

Among these, they can facilitate the development of remote rural economies, thus upgrading the standard of living of the poor.

The main constraint that milk producers seek to overcome by acting collectively is the marketing of their product. The need to be assured of a secure market is a real one. It can be met by dairy farmers cooperatively establishing their own collection system and milk treatment facility in order to convert their perishable primary produce, which requires special and timely attention, into products with longer-keeping quality for marketing purposes.

Most dairy cooperatives adopt either a two—or three-tier system. One village or a group of two or three villages forms the basic unit of the primary cooperative. Only dairy farmers are allowed to enrol as members and they must commit to supplying milk exclusively to the cooperative. While the day-to-day functioning of the cooperative is managed by full-time salaried employees, the committee or board of the cooperative, consisting of only elected members, makes the decisions on the affairs of the cooperative. Primary-level cooperatives bring together members with similar interests at village level to work towards common goals. This system can also identify good leadership talent that would be given a chance to develop through interaction with other community leaders.

A group of primary-level cooperatives forms a union, which can be for a district, region or milkshed area. This is the second tier. The third tier is the unions joining up to form a federation at state or national level, depending on the size and system of administration in the country. The federation has the power to act on such issues as pricing policies, extension, training, control of milk and milk product imports, subsidies and credit.

References

Gupta, P.R., ed., *Dairy India 1983*, New Delhi, Rekha Printers Pvt. Ltd., 1983.

Korten, D.C., "Rural development programming: the learning process approach", Rural Development Committee, *Rural Development Participation Review,* Ithaca, NY, USA, Cornell University, 1981.

Mielke, T., ed., *Oil world annual,* Hamburg, Germany, ISTA Mielke GmbH, 1993.

Zuvakis, C., Jr., *Economic development: an introduction,* New York, NY, USA, St Martin's Press, 1979.

4

Management of Dairy Cow

There are thousands of dairy farms around the world with different farm plans and management strategies. Climate, market conditions, traditions, breeds etc. affect the running and planning of the business. If we try to categorise and generalise the world wide dairy operations, we relatively easily find three main "types". The American type (North and South America) is characterised by large loose-housing operations, total mixed ration feeding (TMR) and relatively many employees. However, dairy farms in Northeast US and parts of Canada differ from the typical American operation. There you find many smaller family farms with either loose-housing or stanchion barns. These operations are quite similar to the European type, which is characterised by relatively small operations where each cow is fed and treated individually.

The third type is mainly found in New Zealand and Australia. The dairy operations in this area are usually run in a very extensive way, relying to a great extent on grazing. Due to the climate and local restrictions, New Zealand and Australian dairy farmers do not rely on concentrate use so much. On top of that, the milk market is de-regulated and the prices follow the world market price (WMP).

When you look closely at all these farm types, they share many underlying principles. Regardless of management strategy, farm size or local climate, the cow has to calf, produce milk, eat, be kept in good health status etc. The goal for any dairy farm should be to maximise profit per unit of milk within farm constraints and, in some parts of the world, within milk quotas. To be able to do so, we need to know how to manage the cow and how different production aspects interact with each other.

Feeding Strategies

As the dairy cow's genetic potential is increasing, feeds and feeding strategies are becoming more and more important. It is well known that the amount of milk to be produced is highly influenced by the amount and quality of the feed given to the cow.

It is also possible to influence the milk composition through the feeding. As the cow normally experiences a shortage of nutrients in early lactation, it is of importance to feed the cow a well balanced diet in order to maximise the dry matter intake. An unbalanced diet increases the risk for metabolic disturbances and weight loss, which have a negative effect on the milk yield. Healthy cows will also make the transition from dry to peak easier.

The cow's largest compartment is the rumen. Together with the reticulum, it has a total volume of approx. 150.200 litres. In these two compartments, there are billions of micro-organisms. They help the cow to digest and utilise the nutrients in the feed. To reach a good feed utilisation, and in the end a high milk yield, the micro-organisms have to have optimal conditions.

Managing Feeding

One of the primary keys to a successful dairy operation is a good nutrition program. Not only is nutrition one of the highest input costs (about 50% of the total costs), but it also controls the results of milk production, reproduction and health. The basis in all feeding programs is the roughage and a common recommendation is that this shall account for 40.60 percent of the total dry matter intake. A cow's intake per day is limited, and it is therefore important to know the dry matter and nutritional content of both the roughage and concentrate. However, the cow's daily dry matter intake is dependent on her stage of lactation. The intake will gradually increase after calving, to reach its peak approx. 6.12 weeks after calving.

When all the feedstuffs (including roughage) have been analysed, we have to make sure that each cow gets what she need in terms of energy, protein, minerals, vitamins and water. To do so there are a number of feeding strategies and systems to choose between. Their common trait is that the roughage intake is controlled through the distribution of concentrates. The more concentrates that are supplied, either separately or in a mix with the roughage, the less roughage the cow will eat.

Ad libitum Feeding

Ad libitum feeding simply means that the animals have free access to the feed and are allowed to eat as much as they want. The roughage is often fed ad libitum regardless of the strategy used for feeding the concentrates (flat rate feeding, challenge feeding and feeding to yield). However, by mixing roughage and concentrates and feeding it ad libitum, the total feed intake for a group of cows can be regulated by means of the degree of concentration of the feed. This strategy is common when using a mixer wagon. A disadvantage with ad libitum feeding is that it is not possible to control each cow's individual feed intake, which increases the risk of both over- and underfeeding.

Flat Rate Feeding

Flat rate feeding is a strategy in which all cows are fed the same quantity of concentrates during the entire (or part of) lactation. The concentrate is restricted on a certain level, while the roughage is fed either ad libitum or in a mix. As described earlier, the cow's energy and nutrient demand varies depending on the stage of lactation. Because of the fixed concentrate ratio, flat rate relies on fat mobilisation. The disadvantages with this strategy are the risk for metabolic disturbances (Ketosis) and the difficulty to reach high peak yields. However, the technique is simple and therefore the investment cost can be relatively low.

Challenge Feeding

While flat rate feeding relies on fat mobilisation, challenge feeding and feeding to yield aims to supply the cow with nutrients that are needed for the actual lactation stage.

Challenge feeding of concentrates takes place in the early lactation, when there is a risk of underfeeding. Every cow is given the maximum ration of concentrates that she can consume, without reducing her roughage intake below approx. 40.60 % of total dry matter intake. This is continued until she reaches peak, four to ten weeks after calving. By using this strategy, each individual cow is given the possibility to 'show' her production potential.

Feeding to Yield

When the peak production is reached and the production starts to decline, there is a risk of overfeeding. By feeding the cow to yield, which means that each cow is fed concentrates individually according to her actual milk yield, the cow's body condition is maintained and the feed is used as efficiently as possible. A successful implementation of this strategy means increased milk production with the feed cost well under control.

Principles of Feeding Systems

The layout of the feeding system and the equipment involved differs depending on the type of barn, stanchion or loose-housing, and the management strategy used on the farm. However, the principles of the feeding systems are generally the same and can be divided into the following four categories: Manual Concentrate Feeding, Computerised Concentrate Feeding, Total Mixed Ration (TMR) and Partly Mixed Ration (PMR).

Manual concentrate feeding

This system is common in stanchion barns but it is also found in loose-housing barns using manually operated in-parlour feeding. In a stanchion barn, the system makes individual feeding of both roughage and concentrate possible. However, in both loose-housing and stanchion barns, the roughage is usually fed ad libitum and the concentrate is fed individually.

Suitable feeding strategies depend on the barn layout. In a stanchion barn it is possible, with a varying grade of precision, to implement challenge feeding and feeding to yield strategies. A loose-housing barn with manual in-parlour feeding is more or less only suitable for flat rate feeding.

A drawback with manual systems is the risk of low accuracy. Practical experience has shown that there often are great differences between the intended ration and actual ration. The diagram below shows that, for example, if an animal should have 4.5 kg, the actual ration could range from 3 kg to 8 kg. This may not only result in costly feed waste and overweight cows, but will also have a negative effect on the milk production. A manual system is also labour demanding and requires more time to implement new feed rations, compared to a computerised system. All this means high feed and labour costs and lost milk revenue opportunities.

Computerised concentrate feeding

Instead of feeding the concentrates manually, they can be dispensed and controlled by a computer. In stanchion barns, a rail suspended feed wagon is normally used. For loose-housing systems, there are two systems to choose from: in-parlour feeding or out-of-parlour feeding (or a combination of both).

The in-parlour feeding is simply a dispenser that can be operated by manual or automatic identification. Out-of-parlour feeders requires automatic identification (transponders). Farms with a long grazing period sometimes prefer to feed all concentrates in the parlour. A disadvantage with this is that large rations cause long eating times and are unhealthy for the rumen. An alternative to this is to use out-of-parlour feeders all year round and keep the cows housed for a couple of hours before letting them out on pasture again.

When out-of-parlour feeding or a rail suspended feed wagon (stanchion barns) is used, challenge feeding and feeding to yield can successfully be used as the feeding strategies. Each cow's individual ration is programmed in a computer and when the cow enters the feed station, she is identified and the correct amount of feed is dispensed. The computer decreases the demand for labour and facilitates the implementation of new feed rations and the record keeping for each cow.

Total Mixed Ration (TMR)

In the Total Mixed Ration feeding system, the concentrates and roughage are mixed in a mixer wagon and usually fed ad libitum. The mix is often dispensed to the cow directly from the mixer wagon, but can also be distributed by band conveyors or with rail suspended feed wagons. The TMR system is most common in large loose-housing herds, but can also be used in stanchion barns.

TMR Feeding

There are mainly two ways to feed the cows with TMR: either feed the same ration to the whole herd (no grouping) or different rations to different groups. In non-grouping systems, only one mix is made and dispensed. The mix is usually composed to suit the high yielding cows, which makes the balance of the ration very important to avoid overweight cows.

A way to improve the precision of the feeding is to use a grouping system. The cows can be grouped according to many criteria: yield, stage of lactation, first calf heifers etc. The drawback with grouping is that much of the rationality of the system is lost compared to the non-grouping system.

Much time must be spent moving the cows from one group to another as the production changes. Group changes also result in milk loss due to social adjustment. In addition, several mixes have to be prepared and dispensed to suit each group. However, if the herd is large enough, it should be possible to run an efficient and convenient TMR system with grouping. Disadvantages with all TMR systems are the relatively high amounts of feed waste and the fact that it is not possible to monitor the individual consumption of concentrates. The cost of feeding and storage equipment is also relatively high.

The main advantage is good rumen health, which results in few metabolic disturbances and possible effects on the milk production. To obtain a well working TMR system and a high dry-matter intake, the ration requires much attention.

Partly Mixed Ration (PMR)

Partly Mixed Ration (PMR) is a feeding regime that combines TMR and individual feeding of concentrates. A mixer wagon is used to mix roughage and some of the concentrates. The concentrate level in the mix is adjusted to fit the lower yielding cows.

The high yielding cows are then fed extra concentrates by out-of-parlour feeders, in-parlour feeders or from a rail suspended feed wagon. The PMR system gives you the possibility to combine the advantages of a mixed ration with computerised feeding.

Milking

To illustrate an individual cow's milk production, we normally plot the yields against time, which gives us the lactation curve illustrated in the picture below. As the diagram shows, the milk yield will rise during the first months after calving, which then is followed by a long period of continuous decline. The shape of the lactation curve will differ from individual to individual and from breed to breed. Feeding and management will also influence the shape and have a significant impact on the total amount of milk produced. Lactation is ideally 305 days, but in practice it is usually more, followed by a two-month dry period prior to the next calving. A cow's milk yield is influenced by many factors, which are described more in detail in the DeLaval booklet Efficient Milking.

Peak Yield

Peak yield is the point where the cow reaches the highest milk production level during the entire lactation. Heifers peak at 70.75% of mature cows and second lactation cows peak at 90% of mature cows. Normally the peak is reached four to ten weeks after calving. The time it takes to reach peak yield varies with many factors, for example breed, nutrition and yield potential. Higher producing animals tend to peak later than low producing ones. A high peak yield normally means a higher total yield. Research shows that each one kilogram increase in peak yield usually means an additional 100.200 kg of milk produced during the actual lactation. Reaching high peak yields requires a very well managed and balanced feeding programme.

Persistency

After the peak, milk production starts to decline by approximately 7.10% per month. The rate of decline is normally measured as the cow's persistency. If a cow's milk production falls by 7% per month after peak, it is more persistent than a cow that loses 10% per month. A general rule of thumb is that higher peak production leads to lower persistency. As for peak yield, the persistency is dependent on feeding and is therefore, to some extent, possible to influence. Persistency differs between cows, but a first lactating cow is normally more persistent than a second or third lactating cow.

Total and Daily Milk Yield

Close monitoring and evaluation of each cow or group are important ingredients when trying to reach a high level of total milk production (high peak and long persistency). However, it is important to remember that a dairy cow is not a stable milk producer. The milk yield differs from day to day and the relative variation can be as high as 6.8 % from one day to another. Cows milked three times or more a day usually have a lower variation than cows milked twice a day.

Managing Milk Yield Recording

The total milk yield is a good indicator of gross income from milk sales, which has a direct effect on the dairy farm's income. With milk yield recording, the dairy farmer can monitor his day to day output and use it to control and monitor the production. Milk yields are also the most interesting variable when calculating the feed rations. Common methods for measuring milk yield are: bulk tank monitoring, monthly milk yield recording, recorder jars/stand alone meters and electronic milk yield registration. The milk yield records can then be used for:

— Calculating and follow-up of the nutrient intake

— Monitoring long-term production changes

— Indicating overall health disturbances

Bulk tank monitoring

Bulk tank monitoring does not enable you to monitor the cows individually, but it can be used as an indicator of the overall herd management. To get the correct total yield, you have to account for the number of fresh, treated and dry cows plus the milk used for own consumption and calves.

Monthly milk yield recording

Monthly recording is usually carried out by a milk yield recording organisation and was originally introduced for breeding purposes. Due to the fact that the recording normally takes place once a month, and each cow's milk production varies day to day, it is not recommended to use this information to manage feeding. When receiving the data, it will be historical and not reflect the current situation.

However, the information can be used for trend analysis and give an idea of the peak and persistency for individual cows, which can be used for culling, breeding and general feeding decisions. As for bulk tank monitoring, the monthly milk yield recording does not give the dairy farmer an early indication of health problems.

Daily milk yield recording

Daily milk yield recording can be considered as one of the most important decision aids for fine tuning the high producing herd. The milk yield can be measured at every milking by using either recorder jars or milk meters. If using recorder jars, all data has to be recorded manually (e.g. on paper or typed into a computer).

With milk meters connected to a processor or computer, you will have an automatic collection of data and a memory capacity that is much more accurate and efficient than the milkers. The processor contains a database where all individual milk yields and other relevant data are stored on a daily basis. This data provides the manager with exact and timely information of a cow's or group's production on a day to day basis, which can be used to:

- Calculate and evaluate feed rations for individual cows or groups to maximise the milk production and avoid over and/or under feeding.
- Identify cows in heat. Research shows that decline in morning milk might indicate oestrus.
- Identify cows with potential health problems early. For example, ketosis is related to a gradual decrease in milk production before it can be diagnosed.
- Calculate the actual lactation curve instead of estimating.
- Group cows by yield, which results in faster milking.
- Evaluate long-term milk production for individual cows to plan breeding and culling.

Reproduction

Reproduction is a necessary and important part of milk production. Without regular calvings in the long run, there will be difficulties in producing the desired amount of milk. It is also important to produce sufficient heifer calves as replacement animals and allow herd size to be maintained or expanded.

Today it is most common to use artificial insemination (AI) instead of a bull. AI increases the control over the breeding and enables the dairy farmer to use sperm from all over the world, which increases the genetic gain. However, at the same time, it increases the need for good planning and well structured working routines.

Another way to get a cow pregnant is to use the embryo transfer technique. So far this is relatively uncommon, but it is expected to increase in importance. The main advantage with embryo transfer is that it is possible to generate more calves from a good cow than it is with AI.

The oestrus cycle

As long as a cow or heifer is not pregnant she will normally have a 21-day oestrus cycle. The length of the oestrus cycle may vary but it usually ranges from about 17 to 24 days. A heifer's oestrus cycle is normally slightly shorter than a cow's. The cycle will continue until the cow is pregnant. After calving, cows normally undergo a 20 to 30-day period when oestrus cycles do not occur.

The oestrus cycle is controlled by a complex system involving different hormones produced in the brain and ovary. The picture below shows a simplified picture of how two of these hormones, oestrogen and progesterone vary depending on where in the cycle the cow is.

Some cows do not follow the normal oestrus cycle. For example, a cow can be unoestrus, which means her ovaries do not function with the regular 17 to 24-day cycle and are therefore not observed in heat. Other cows may suffer from ovarian cysts. These cows will show heat at very short intervals and the period that they are in heat will last three to four days.

Reproductive management

Research has shown that suboptimal calving interval causes large economic losses to dairy farmers, second in importance only to mastitis. Milking cows in late lactation are less profitable due to the decline in production. A long calving interval means milking less profitable cows due to the decline in production, fewer calves and too many cows with low feed conversion efficiencies. Therefore, successful reproductive management has a significant impact on the herd's overall performance and the net income.

Well managed reproduction also reduces the risk of expensive involuntary culling. Information from UK and USA shows that the cost for a one day extended calving interval ranges from £1.5 to £3 per cow. The time of the calving interval that is possible to influence by management is the open days, which are determined by the voluntary waiting period (VWP) and the breeding window (BW). A very common reason for undesired long calving intervals are missed heats. With improved heat detection rate (HDR) and conception rate (CR), through better management and improved timing of inseminations, it is possible to obtain a significantly shorter calving interval.

Detecting heat

The most sexually intensive period of the oestrus cycle is during standing heat, which lasts for approximately 18 hours. In loose-housing herds, this period is indicated by the cow in heat standing immobile when mounted by another cow or bull. Other signs of heat are:

— Bellowing

— Increased activity

— Walking the fence line

— Licking/Sniffing

— Swelling and reddening of the vulva

— Mounting other cows

— Lower milk yield
— Reduced feed intake

The duration of heat varies from animal to animal, but approximately 10 to 12 hours after the end of standing heat, the egg is released (ovulation) and the heat ends.

Manual heat detection

Manual heat detection relies on manual observations in the barn. The cows and heifers shall be observed for oestrus two to three times per day and all observed heats be recorded whether the animal is bred or not. The records provide the manager with information to anticipate future heats, which will make it easier to distinguish if a cow is in heat or not. Most mounting occurs between 6 p.m. and 6 a.m., and it is therefore worth trying to check for heat during these hours. To facilitate the planning and record keeping, a cow calendar is often used. This can be either manual or computerised.

The disadvantage with manual detection is that it is very time demanding and requires people with the ability to observe the right signs. This is particularly important when there are no distinct signs of heat. By using progesterone testing or other tools on the market, the detection rate can be improved.

Automatic heat detection

Another way to identify cows in heat is to monitor their activity. During heat the activity can increase up to 8 times compared to the normal level. The activity can automatically be recorded by using activity meters attached to the neck or leg. By comparing the activity with the last observed heat, actual milk yield and feed consumption, a reliable indicator of heat is obtained. The automation will generate a significant time saving and improved calving interval through better heat detection.

Timing of insemination

With artificial insemination, the timing of the insemination becomes important. The optimum time for insemination depends on when the ovulation occurs in relation to the heat and for how long the sperm is viable. Most sperm remain viable for about 24 hours. The ovum's life is only about four hours and is the most critical. It is therefore preferable that viable sperm is present in the salpin during ovulation.

There are mainly two rules for the timing of insemination. Traditionally the a.m.-p.m. rule was the one dairy producers followed. This rule dictates that cows and heifers first observed in heat in the morning should be bred late in the afternoon. Likewise, cows and heifers first observed in heat in the afternoon should be bred the following morning.

This is still a good rule, but many producers have now successfully gone to once-a-day insemination. This rule dictates that cows and heifers first observed in oestrus in the afternoon or the following morning, should be bred late that morning.

Herd Health

Disturbances and diseases will always occur in a herd, but in order to limit the economic loss, it is important to keep them under control. Good health and health management have a significant effect on a dairy farm's net result. As the number of different disturbances is great, only the most common ones will be covered here.

Common Diseases and Disturbances

— *Retained placenta*: This is a condition where the foetal membranes are not completely expelled. Retained placenta increases the risk for a whole host of other health problems (e.g., uterus infection, ketosis, and displaced abomasum). Usually a retained placenta is associated with difficult calving, a mineral imbalance (e.g., potassium, calcium) prior to calving, or overconditioning.

— *Milk fever*: Milk fever occurs when there is a shortage of calcium in the blood. This also usually results from a mineral imbalance (e.g., potassium, calcium) prior to calving. The effect of milk fever normally starts with reduced feed intake, followed by difficulties for the cow to move. Eventually the cow becomes paralysed and, if not treated, may die within a couple of hours. However, treatment with calcium usually has a good effect and in most cases, the cow will survive.

— *Ketosis*: Ketosis occurs when cows begin to milk but have a shortage of energy. These cows will use fat to support milk production, but their liver cannot convert the fat to energy fast enough. Ketone bodies accumulate in the blood causing ketosis. Treatments generally consist of rapidly digested sources of glucose such as propylene glycol or molasses.

— *Left-Side Displacement of Abomasum (LDA):* This may occur in relation to calving. The fourth stomach (abomasum) migrates from the right side of the cow to the left side. It is thought that this occurs with low fibre levels in the diet, physical stress such as slippery floors, and secondarily to other problems such as ketosis.

— *Acidosis*: There are a few health problems directly associated with feeding, but high levels of concentrates might cause acidosis. The rumen pH drops below 6.0 for a long period of time causing numerous problems. One result of acidosis is hoof problems. Cows that have acidosis may experience hoof ulcers and a form of founder.

— *Mastitis*: Mastitis is the most common and costly disease in dairy herds. It is an inflammation in the mammary gland which can be caused by bacterial infections or trauma. Cows with clinical mastitis are relatively easy to detect for the farmer. The symptoms are clotting and discoloration of the milk, and the gland becomes hard, red or swollen and in severe cases the cow has fever and loss of appetite. Cows with subclinical mastitis will have no outward signs of mastitis but their mammary gland will still be infected and the somatic cell count will be high.

— *Lameness*: Hoof problems or lameness is characterised by damage to sensitive laminae of the hooves and it is reported to be the third most common cause of culling behind reproduction and mastitis. Hard and slippery floors, decreased exercise and metabolic disturbances (especially acidosis) negatively affect the occurrence of hoof problems. Estimates are that cattle that become lame and are not tended to, can experience a 20% loss in milk production.

Managing herd health

The economic losses due to health disturbances can be attributed to one or more of the following factors:

— Less efficient production and higher veterinary costs

— Reduced slaughter value and idle production factors

— Lost future income

Lost future income occurs when the cows have to be replaced (culled) before reaching their economically optimal age. This varies from individual to individual, but it is basically the time when it is more profitable to replace a cow than to keep her. Culling is usually the eventual outcome of health problems and culling rates are normally found to be between 25.35%. Cows that die on the farm are part of this culling rate and should represent less than 3% of the total herd.

The reduction of losses attributed to diseases and the promotion of positive welfare and health should be one of the main goals in dairy herd management. A helpful tool for maintaining healthy cows is body condition scoring (BCS). Other important aspects that might affect health, if not looked after and maintained, are barn environment, milking routines and milking equipment.

Body Condition Scoring (BCS)

Body condition scoring (BCS) can be used to troubleshoot

problems and improve the health and productivity of the dairy herd. Overweight cows (overconditioned) are more susceptible to metabolic problems, infections and noninfectious health problems. Research shows that overweight cows are more likely to be affected by, for example, mastitis, retained placenta, ketosis and lameness. They are also more likely to have difficulties at calving. Thinness (underconditioning) can lower the milk production and fat content because of insufficient energy and protein reserves.

The body condition scoring system uses a 'one to five' scale with one representing very thin cows and five very overweight cows. An 'ideal' cow has a body condition score of about 3.5, but the system is designed to have cows at certain stages of lactation at certain body conditions (table below).

Score	Condition
1	Skin and bones
2-2.5	Severe negative energy balance on cow in early lactation. Risk of production loss.
2.5-3	High producer in early lactation.
3-3.5	Milking cow in good nutrient balance.
3.5-4	Late lactation and dry cow in good condition.
4	Overconditioned. Potential calving problems if dry.
5	Severely overconditioned. Risk of fat cow syndrome.

Dry cows should neither gain nor lose weight and their body condition score should be around 3.5. If they have too much fat (>4.0) they tend to have many problems after calving. The reason is that all of this body fat must move through the liver before it can be used for milk production. This is not bad if they use the fat slowly, but if they begin to lose weight quickly, fat will accumulate in the liver and a condition known as 'fatty liver' develops. These cows usually perform poorly and are more prone to die. After calving, cows should lose less than one point before they begin to gain weight again. Cows that lose more than one point tend to have more reproductive problems.

From peak milk production to dry off, cows should gain back the body condition that was lost before the peak was reached. It is common for cows that have problems (e.g., ketosis, hoof problems, mastitis etc.) to lose one body condition score within the first two weeks of lactation. Ideally, the body condition score of a cow should be taken whenever she is handled. At least body condition scores should be taken at freshening, breeding and dry off. If over- or underconditioning occurs, the feeding strategy for an individual cow or group has to be evaluated and feed rations may have to be recalculated and corrected to satisfy the cow's demand.

Preventing and Controlling Mastitis

Mastitis needs extra attention, as it is the most common disease among dairy herds. It is not unusual that up to 40% of the herd is infected by clinical (with visible symptoms) or subclinical (often without symptoms) mastitis. With good management routines, there are great opportunities to reduce the number of cases. This has a direct effect on production levels and costs. For example, if a farm is able to reduce the somatic cell count from 200 000 to 100 000, it means an increased average production of 0.7 kg per cow and day. In a 100 cow herd, this gives an additional milk production of 0.7 X 365 X 100 = 25 550 kg of milk per year.

The management must focus on actions that prevent new infections and reduces the duration of each infection. This includes proper maintenance of bedding and stalls, hygiene procedures during milking and frequent checking of the milking equipment. Dairy cows require a comfortable, clean, dry and draft-free environment. Comfort can include everything from feed space and walking surfaces to air and resting areas. However, even if the physical facilities are well planned, the daily work routines are of great importance to prevent as many health problems as possible. This includes not only mastitis, but also hoof problems (lameness). The way to avoid many of these cases is to regularly trim the hoofs and make sure that the surfaces in the barn are dry and free from mud.

Quality of Milk

Milk consumers require healthy and safe milk that is free from contamination, unpleasant smell etc. Processing also demands for milk to meet certain standards. For example, cheese making is very sensitive to spores and antibiotics. The requirements on the quality has increased over the time and will continue to do so. To meet this, more and more dairies put pressure on the milk producers. This is normally reflected in the payment scheme, where producers that deliver milk that does not meet the required standards have to pay penalties. There are also examples of processors paying bonuses to producers that meet the highest standards. As this makes the quality of the milk a determinant of the producer's final price, it has to be considered in the daily routines.

The dairies use different parameters to check the quality of the milk. The exact level for each parameter varies from company to company. Therefore it is worth taking a closer look at the demands and standards you have to meet. Examples of tests carried out on delivered milk are:

- — Fat and protein content
- — Bacteria count (Bactoscan)
- — Smell and taste
- — Residues from antibiotics and other Medicines
- — Freezing point (detects water content)
- — Somatic Cell Count
- — Spores (especially important for cheese making)

Managing Milk Quality

There are many factors that affect milk quality and some are easier to control than others. With well structured daily routines and an awareness among the personnel, it is possible to control and improve the situation considerably.

References

Heinrichs, A.J., "Drying off the dairy cow", In: *Dairy Management Manual*, A Western Regional Extension Publication, 1990.

________________., "Dry cow management", In: *Dairy Management Manual*, A Western Regional Extension Publication, 1990.

Smith, R.D., L.E. Chase and C.J. Sniffen, "Managing the dry cow for top production, reproduction and health", *In: Dairy Management Manual*, A Western Regional Extension Publication, 1990.

5

Dairy Cow Breeding

Raising cattle for milk production requires close attention to breeding them so that they produce the most milk possible. This will require some planning, close observation and recording of matings and regular recording of the daily milk production of all of the cows in the herd.

There are four important considerations in breeding for milk production.

- — Selecting the best cows
- — Selecting the best bull
- — Calf castration and selecting the best replacement bulls and heifers.
- — Achieving the best breeding efficiency.

Selecting the Cow

All cows can produce some milk. This is usually too little to be worth collecting for sale, although it may be enough for family consumption. There are some dairy bred cows and cross bred dairy cows (for instance Friesian and Shorthorn breeds) to be found in smallholder beef herds, this are often the descendants of cattle introduced by church missions for their own dairy herds. These can produce more milk than average cattle. They should be selected as the nucleus for starting a dairy herd.

If you have this cattle type already, you can either begin to produce milk from them by feeding them better and using them as the beginning of a dairy herd and keeping them until they are replaced by better milk producing heifers or cows. Then these older, starting animals should be sold. There is a classical description of the triangle-shaped dairy cow which can be used for selection if there are enough dairy heifers around for you to choose. The best heifers and cows for you to start with are those which are quiet in temperament and which have some evidence of dairy blood. If there are none of these, start with quiet local animals with a history of successful breeding, and use a dairy bull over them to produce upgraded offspring.

Thus the long term plan should be to replace the original cows with heifers better bred for milk production. Such heifers can be produced from these cows by breeding them to a good, dairy type bull. To start a dairy herd, the first cows should have these characteristics:

- — be of obvious dairy stock i.e. crossbreeds,
- — preferably already having calved and therefore demonstrating that they can produce calves
- — have a good udder and teats
- — have a quiet and handleable temperament
- — be as close as possible in shape and appearance to the cow in the photograph above.

Selecting the Bull

The bulls which are generally available on most small farms are unlikely to be useful for upgrading your present cows for dairy production. The safest and most efficient way to obtain a breeding bull is to sell (or castrate) all of the existing bulls on your farm, and replace them with a good dairy type bull from a respected private, school, church or government breeding farm, If you can not get a pure-bred, you should use one which has at least half dairy blood. In this way you can be assured that his daughters will be superior in milk production to any animals that you already have, or that can be fathered by any other bulls in your neighbourhood.

You should be able to produce two calves fathered by this bull from any of your mature cows before he is sold, when he should be sold or traded to another farmer after three years to prevent him mating his own daughters. Keeping him longer so that he sires offspring from his daughters may result in those offspring having defects which make them unhealthy or unsuitable for production. if you have a choice of bull breed from the farms, it would be better to alternate the breed each time you buy a new bull, because same-breed bulls from the small herds on the farms in Island countries may be too closely related for you to buy unrelated bulls of the same breed every time.

Besides increasing the milk production characteristics of their daughters, most dairy type bulls will also throw excellent beef type calves. In addition to gaining the advantage of milk production from the daughters, a dairy bull will also increase the carcass value of the cattle you sell for beef.

In selecting and using a dairy breeding bull, you should:

— select a good bull from a known source (a government breeding farm),
— make sure he is sound, not lame and can serve properly,
— castrate all other bulls you have,
— prevent outside bulls from coming onto your farm and mating with your cows,
— do not allow the bull to mate with his daughters and replace him after about three years,
— select a replacement bull of a different breed [if available],

Organising the Mating

You can num the bull with the cows and heifers all year round. This will simplify' your management, but:

— your cows will produce calves all year round and some will calve during the season when the feed is of poor quality and so will probably not milk well without being fed concentrates, and,

— you will probably not know when each cow was mated, so you will not know when she will calve, nor will you know whether she is pregnant or not until she is almost ready to calve. This means that if there is something wrong with her (or the bull) you might not know until almost a year later,

If you are on the farm all the time or you have good helpers who are observant and interested, it is better to keep the bull confined in a small bull paddock, and detect the cows which are ready to be mated by observing the cows and heifers every day to see if they show signs of oestrus. Cows which are seen to be in oestrus can then be taken to the bull and the mating observed and recorded. The advantages of this breeding system are:

— you can take good care of the bull and make sure he doesn't become injured in the field. He is also quieter and much more easy to manage,

— you will know if a cow is showing breeding activity and you know the breeding dates if she is joined. From this you can predict her calving date, thus,

— you can keep accurate records,

— you can anticipate her next mating time and be ready to join her again if she does not conceive the first time,

— you can quickly tell if you have a problem in the herd if cows continually return to the bull for several months without becoming pregnant,

— you can quickly identify cows or heifers which have reproductive problems

The disadvantage of this system is:

— you need a separate pen for the bull

— you must observe the cows and heifers every day

Signs of Oestrus

If your cattle are quiet and easy to handle (and all dairy cows and bulls on small farms should be), you can take the cow to the bull and leave her loose in the bull paddock, or you can tie her up to a tree with a rope or halter, and lead the bull to her and allow him to mount. You will know she is ready to be mated when she shows:

- bellowing and perhaps walking up and down a fence line,
- streams of clear mucus coming from the vulva,
- mounting or mounted by other cows,
- if she is running with other cows, her tail may be slightly raised and the hair on the top of the base of the tail may be roughened and standing up,
- if she is milking, her milk production may suddenly fall a little.

Heifers may not show these oestrus signs as strongly as cows. So it is preferable to keep heifers and cows separate, and introduce the bull to the heifer paddock every day after the time when you think they are old and big enough to mate. You should then record any matings that occur. Often if the heifers are old and big enough to mate but for some reason are not showing signs of oestrus, the presence or sight of a bull will cause them to come into oestrus a few days later.

So keeping the heifers within sight of the bull can be useful, provided your fences are good. Keeping heifers and cows separate until the heifers enter the milking herd is an important principle, as mature cows will dominate the younger ones, often leading to a loss of weight and delayed breeding activity. Join heifers to the main herd of mature cows only after they have calved.

Oestrus occurs every 18-21 days and lasts for about 18 hours. A cow or heifer will continue to come into oestrus every 18-21 days until she is successfully mated and becomes pregnant. If you see a cow in oestrus in the morning, it is best to have her mated to the bull at that time, and again in the afternoon of the same day. Once a cow becomes pregnant she will not usually come again into oestrus until after she has calved. A very small percentage of heifers will come into oestrus and accept the bull even though they are already pregnant, but this characteristic is not particularly important, particularly if you keep good records.

Goals of Breeding

Cattle breeding is a relatively simple endeavour. The only difficult part is to keep it simple". In breeding dairy cattle, certain objectives must be set out. These include:

- to increase milk production
- to breed better functional characteristics that will allow the cow to produce more milk over a longer period of time.

It is an accepted principle that the more characteristics for which one selects, the less progress is made. Therefore, in line with the above quotation, it would be advisable to restrict selection criteria to the following:

- milk yield
- fat and protein yield
- feet and legs
- udders
- capacity
- "dairyness"

Each will be analysed separately. Collectively, selection for these six characteristics would go a long way to ensuring more productive dairy cows in South Africa.

Milk Yield

A dairy cow produces milk. Simple as that statement might seem, it is imperative to select primarily for an increase in milk production. It is of no use at all to breed a fantastic looking cow which produces no milk. In order to achieve this, the farmer must select bulls that are positive (on whatever ratings) for milk production.

An average bull in any category is likely to breed average-yielding daughters, *i.e.* most of them will be average cows. Very few average bulls will breed exceptional daughters. The whole breeding programme is negated by selecting bulls that are negative for milk production. After all, a dairy farmer theoretically wants to run fewer cows yielding more milk. It is inconceivable to think of using a minus milk index *i.e.* below average index bull.

Fat and Protein Yield

American research has shown clearly that selecting for milk yield only also increases the total fat and protein yield. However, the percentages (or deviations) of fat and protein from the average would remain more or less the same. The same can be said if one selects only for fat and/or protein yield, *i.e.*, a corresponding lift in total production will occur. However, selection entirely on percentages, *i.e.*, +0.00 % fat, or +0.00 % protein, or both, will decrease milk yield by as much as 300 kg. There will also be a significant decrease in component yield.

Feet and Legs

South African dairy farmers have always boasted how tough and extensive their farming conditions are. Our cows may have to walk long distances to and from their feed, unlike cows elsewhere in the world. It is therefore imperative that the cow has good feet and strong legs.

What are good feet and legs? One is looking for a hind leg which is slightly sickle-shaped (from the side view) with a steep (strongly-attached) pastern. Only bulls breeding good feet and legs should be used in a dairy herd.

Cows with weak pasterns are a curse to any dairy farmer, because the cow then walks on the soft part of her "heel" (actually on flesh), and not on her hoof as she is supposed to.

On can conclude that she will only be able to walk with great difficulty, and will suffer from numerous infections.

Cows with straight, or post hocks, are to be avoided, because they are highly heritable (passed on from generation to generation), and cause cows to walk with an abnormal stiff-legged gait.

Any cow which is unable to stand up and/or walk with ease is useless, even if she has the most perfect udder in the world. Functional legs are very necessary on a dairy cow. Cows which are able to walk properly are long lasting, "no hassle" cows which are definite economic units in the dairy enterprise.

Udders

The udder must be pliable, silky in texture and sack-like in nature. Ideally, the udder, when viewed from the side, ideally, should not hang below the cow's hock. The single most important part of the udder is the central, or median, suspensory ligament. This must be extremely strong and well attached. It is an accepted fact that an udder with an excellent central ligament is a long-lasting one.

Teat placement is next in importance. Ideally, the front teats should be even and centrally placed on each quarter of the udder. Simple as this may seem, many cows in South Africa suffer from the inherent problem of wide front teats. Teat size (over- and undersized teats should be avoided), shape and placement are highly heritable. Great emphasis must be placed on a bull that breeds improvement in all udder traits, or characteristics, especially a strong central ligament and acceptable teat placement and size.

Body capacity

Viewed from the side, a cow with a deep, long body with wide, well-sprung ribs is said to have a large body capacity. Large body capacity is associated with ample space for the rumen and digestive system, and this, in turn, is associated with superior milk production. It should be obvious, but still needs to be stated, that a dairy cow with little body capacity cannot be a great milk producer. This is because the gut size is limited by the capacity of the abdominal cavity.

Besides a deep body, what are the other pointers of capacity? These are:

— A broad muzzle ("shovel-nose" the Americans call it). A broad, strong muzzle implies the ability to get the food into her mouth and to masticate (chew her cud) effectively

— Width between the fore legs. This shows whether, or not, there is plenty of room for the vital organs situated between the shoulders and front legs. Cows with a narrow chest are not normally good producers

— Width of rib. If at least two fingers, can be placed between the ribs of a dairy cow, she is said to have a fair degree of capacity. Ideally, in any cow, three flattened fingers would indicate great capacity.

These pointers are quite acceptable rules-of-thumb to measure body capacity. The bulls ability to breed capacity should be considered in any breeding programme. However, milk yield, fat and protein content, feet and legs, and udders, are most important, whereas with body capacity a certain amount of leniency is allowed.

Dairyness

Dairyness is a subjective evaluation made on dairy cows. It is extremely difficult to measure, and equally difficult to describe. It is not more important than the preceding five criteria. Dairy cows are refined animals which produce milk. Beef cattle produce beef, and are solid and well muscled. We hardly expect beef cows to produce the same volumes of milk as do dairy cows.

What, then, is the refinement we want to see in dairy cows? The following are good pointers:

— Refinement can be related to sharpness across the shoulders (or crops) instead of being broad (thick) and beefy.

- — It can also be related to flatness of bone, seen especially on the inner thigh where the bone should be flat and "clean" rather than strong and coarse.
- — A thin, fine tail instead of a thick, robust and coarse tail.

These are subjective judgements, arising from the observations of practical dairymen, rather than scientific facts.

Bulls which breed high-producing daughters, tend to breed "dairy" looking daughters, with good feet and legs, a functional udder, great capacity and is sharp over the chines and has a clean, flat bone on the inner thigh. Emphasis should therefore be placed on bulls breeding dairyness (or sharpness).

Heritability

The degree to which a bull is able to influence various characteristics in his progeny genetically is measured by heritability. Different characteristics (traits) have different heritabilities. This obviously influences the development of the breeding policy because faster genetic progress can be achieved for traits which are higher in heritability, compared to traits which are lower in heritability. It is difficult to make much genetic progress through selection and mating unless a trait has a heritability of .10 or higher.

Breeding Policy

Registered Dairy Cows

If the farmer belongs to a breed society and has his cows registered with that society, that enables him to obtain semen from outside the borders of South Africa. In most herds, complimentary mating is practised where the short comings of each cow, *e.g.* high pins, shallow heels, faults which were pronounced in the daughters of S-W-D VALIANT, are matched with a bull which will improve and compliment the daughters characteristics. In this case a bull which sires level or sloping rumps, steep heels, *e.g.* an ELEVATION son, would compliment the characteristics of the cow to be mated.

Grade dairy cows

Almost all dairy farmers keep records and so obviously, the farmers with grade dairy cows would avoid using a bull on his daughters. Such inbreeding would most definitely be undesirable. If however, there is blanket mating of the herd with one or two bulls, the following are useful guidelines:

- — 100 units of semen will yield 20 milking daughters of a bull.
- — 2 bulls should be used per 100 cows, and these bulls should be sold after 3 years before they came back to cover their daughters

In any cattle herd, between 20 and 30 daughters of any bull can give the breeder a realistic impression of the breeding value of that sire. Far too many dairy farmers use each and every bull, resulting in between 1 and 5 daughters per bull. With such a small sample, no positive evaluation can be made regarding the impact, or complementarity, of the bulls used in the herd.

It is vitally important to make a sound selection and to use such a bull extensively (mated to at least 50 cows) which would then enable the breeder to gauge his worth. Obviously if the bull is successful, he could be used extensively to produce more, desirable daughters. If he is unavailable, then a bull of similar, if not identical breeding, can often fill the gap.

Semen pricing

Remember that the most expensive semen is not necessarily from the best bull, especially in the case of overseas bulls.

Two options are available to the dairy farmers:

- — Use the best bull across the herd, *viz* if there are 300 cows to be bred, use the best bulls available (50 cows per bull).
- — Alternatively, and somewhat more economically, you can divide the herd into three:
  - — *Top producers*: use the best bulls on them

- *Middle producers*: use the best reasonably priced bulls on them, not forsaking milk production
- *Low producers*: these should be bred to beef bulls/ semen, thereby ensuring better management of fewer replacement heifers, out of the top section of the dairy herd, and a gradual elimination of the weaker cows and their progeny out of the herd. It is all very well to say, use the best bulls across the herd, but in many instances cash flow and other financial constraints prevent this management/ breeding decision.

Corrective Mating

Many breeders, including the old masters of the art, believe in corrective mating. Under corrective mating, cows which may be highly productive yet have some objectionable characteristics are mated to a bull that is especially outstanding in the trait in which the cow is inferior. The object of this article is not to prove or disprove this theory. What is important is that the six essentials discussed are adhered to in any sire selection. If this is done, then it would be unnecessary to apply corrective mating down to the last minute detail.

What is important is the complementarity of bloodlines (nicking ability), *e.g.* CHIEF X BELL or VALIANT X ELEVATION to name a few. Do not inbreed. Ensure that the selected sire compliments the breeding of the cows he is to be mated to. Ask your breeding consultant for further advice.

Selecting a Bull from a Dairy Semen Sire Directory

Such a directory ultimately puts together all the above selection criteria, allowing an effective decision in the choice of a bull. Make sure the selection is a good one, using the above guidelines, as the resulting progeny will only be in production in five years time. Poor decisions can force you out of business.

Calving Ease

A rule of thumb in this regard is that one calving out of ten is difficult. If this figure falls to one difficult calving out of every three, then there is a large problem. Therefore the ideal easy calving figure is between 5% and 12% whilst bulls with between 15% and 25% should be avoided. While calving difficulty is often closely related tofeeding levels, some bulls are problematic.

Selection of Dairy Cows for Freeding

All cattle breeds, and some goat breeds, have the ability to supply milk surplus to the needs of their young. If one is keeping just one or two cows for purely household purposes then it is not essential to keep dairy cows, since even beef cows can be milked. However, for commercial milk production, consistently high yields are essential if a dairy enterprise is to retain its viability. In the major dairy countries of the world, such as Israel, the production tendency was towards high yields within a short lifetime. However, certain problems arose, as a result of which the trend throughout the world has changed towards selection for high yields with a good butterfat and protein composition, essential under the emphasis for high component pricing for protein, over a long lifetime.

In selecting a breed, the farmer will do well to examine his farming enterprise with care and to choose a breed that suits his requirements. The availability of both cows and bulls is an important consideration in this choice. Jerseys are far more heat resistant than Holstein-Frieslands and are also better foragers. Therefore, Jerseys are more suited to hot areas such as the Transvaal Lowveld, and to more extensive dairying. Ayrshires are also said to be good foragers but they are more sensitive to bad stockmanship than are Holstein-Frieslands or Jerseys.

Animals can usually adapt to a new environment, but this can be a lengthy process. Buying animals from an area with similar climatic conditions, preferably close afield, is therefore a commendable practice.

Artificial Insemination in Dairy Cattle

Artificial insemination (AI) is a process by which sperm are collected from the male, processed, stored and artificially introduced into the female reproductive tract for the purpose of conception. AI has become one of the most important techniques ever devised for the genetic improvement of farm animals. It has been most widely used for breeding dairy cattle and has made bulls of high genetic merit available to all.

The history of AI is interesting. Old Arabian documents dated around 1322 A.D. indicate that an Arab chieftain wanted to mate his prize mare to an outstanding stallion owned by an enemy. He introduced a wand of cotton into the mare's reproductive tract, then used it to sexually excite the stallion causing him to ejaculate. The semen was introduced into the mare resulting in conception.

Anthony van Leeuwenhook, inventor of the microscope, first observed human spermatozoa under magnification. This finding led to further research. Spallanzani is usually considered the inventor of AI. His scientific reports of 1780 indicate successful use of AI in dogs.

In 1899, Ivanoff of Russia pioneered AI research in birds, horses, cattle and sheep. He was apparently the first to successfully inseminate cattle artificially. Mass breeding of cows via AI was first accomplished in Russia, where 19,800 cows were bred in 1931. Denmark was first to establish an AI cooperative association in 1936. E.J. Perry of New Jersey visited the AI facilities in Denmark and established the first United States AI cooperative in 1938 at the New Jersey State College of Agriculture.

The AI industry has grown tremendously in the United States since its beginning. In 1970, USDA reported that 7,344,420 dairy females were bred artificially, 46% of the female dairy cattle population.

Advantages and Disadvantages of AI

The greatest advantage of AI is that it makes possible maximum use of superior sires. Natural service would probably limit the use of one bull to less than 100 matings per year. In 1968, AI usage enabled one dairy sire to provide semen for more than 60,000 services. Exposure of sires to infectious genital diseases is prevented by use of AI which reduces the danger of spreading such diseases. Time required to establish a reliable proof on young bulls is reduced through AI use. Other advantages include early detection of infertile bulls, use of old or crippled bulls and elimination of danger from handling unruly bulls.

There are a few disadvantages of AI which can be overcome through proper management. A human detection of heat is required. Success or failure of AI depends on how well this task is performed. AI requires more labor, facilities and managerial skill than natural service. Proper implementation of AI requires special training, skill and practice. Utilisation of few sires, as occurs with AI, can reduce the genetic base. The AI industry and dairy cattle breeders should make every effort to sample as many young sires as possible.

Collection, Extension and Storage of Semen

One obvious factor which determines degree of success of AI is the quality of the semen used. Much has been learned about factors affecting semen quality and methods of evaluation and means of maintaining quality through lengthy storage. The commercial AI industry has a tremendous responsibility to sell only high quality semen. Unquestionably, they are fulfilling this obligation.

Collection of Semen

Several methods of obtaining semen have been developed. The artificial vagina method is most widely used today for the collection of bull semen. The bull is allowed to mount a teaser cow and ejaculates when the penis is directed into the artificial vagina. The artificial vagina consists of a firm cylindrical tube with a thin-walled rubber lining. The jacket formed is filled with warm water. A rubber funnel connected to a collection receptacle is attached to one end of the cylinder. When the jacket is properly filled and the artificial vagina lubricated and properly applied, this method of semen collection is highly successful.

Cleanliness must be practiced to avoid contamination and deterioration of semen quality. Proper and careful treatment of the bull is essential to bring about adequate precollection stimulation which will increase quantity and quality of semen obtained.

Obviously the collection of semen from a bull is a specialised skill and should be attempted only by those with the proper equipment, training and experience. Adequate facilities for controlling the bull and teaser animal must be maintained so that danger of injury to personnel as well as the animals is minimised.

Semen Extension

The main reason for extending (diluting) semen is to increase the number of females serviced from one ejaculation. A normal ejaculate from a dairy bull will contain 5 to 10 billion sperm which can be used to inseminate 300 to 1000 cows if fully extended. There are several good semen extenders. Those made from egg yolk or pasteurised, homogenised milk are two of the most widely used. A good extender not only adds volume to the ejaculate but favours sperm survival and longevity. Dilution rate depends on quality of the ejaculate—number of sperm cells, percent alive and mobility. As few as 12 million sperm per insemination have given good conception rates.

Penicillin and streptomycin are added to semen extenders. These antibiotics inhibit bacterial growth and reduce danger of spreading diseases such as vibriosis.

Semen Storage

The discovery that bull semen could be successfully frozen

and stored for indefinite periods has revolutionised AI in cattle. In 1949, British scientists discovered that addition of glycerol to the semen extender improved resistance of sperm to freezing. Glycerol acts to remove water from the sperm cell prior to freezing and prevents the formation of cellular ice crystals which would damage the sperm. There are two methods of freezing and storing semen: dry ice and alcohol (-100 degrees F) and liquid nitrogen (-320 degrees F). Liquid nitrogen is preferred because there is no evidence of fertility deterioration with age. Fertility gradually declines in semen stored in dry ice-alcohol.

Frozen semen can be stored indefinitely if proper temperature is maintained. A recent report told of a calf born from frozen semen stored for 16 years. Fresh, liquid semen can be successfully stored for 1 to 4 days at 40 degrees F. Semen is usually stored in glass ampules. Other methods appear promising, particularly the French-straw. Several AI organisations have gone to this method exclusively.

Artificial coloring is frequently added to semen extenders in order to distinguish one breed from another. Complete identification of the bull is required on each individual semen container.

Artificial Insemination Techniques

The technique of inseminating a cow is a skill requiring adequate knowledge, experience and patience. Improper AI techniques can negate all other efforts to obtain conception. Semen must be deposited within the tract of the cow at the best location and at the best time to obtain acceptable conception rates.

Early methods of AI involved deposition of the semen in the vagina, as would occur in natural mating. Those methods are not satisfactory. Fertility is low and greater numbers of sperm are required. Another method which gained popularity was the "speculum" method. This method is easily learned, but proper cleaning and sterilising of the equipment is necessary, making it more impractical to inseminate than with the rectovaginal technique which is the most widely used AI method today.

In the rectovaginal technique a sterile, disposable catheter containing the thawed semen is inserted into the vagina and then guided into the cervix by means of a gloved hand in the rectum. The inseminating catheter is passed through the spiral folds of the cow's cervix into the uterus. Part of the semen is deposited just inside the uterus and the remainder in the cervix as the catheter is withdrawn. Expulsion of the semen should be accomplished slowly and deliberately to avoid excessive sperm losses in the catheter.

The body of the uterus is short; therefore, care should be taken not to penetrate too deeply which might cause physical injury. In animals previously inseminated, the catheter should not be forced through the cervix since pregnancy is a possibility. Since research data show little variation in conception rates when semen is placed in the cervix, uterine body or uterine horns, some people recommend incomplete penetration of the cervical canal and deposition of semen in the cervix.

The rectovaginal technique is more difficult to learn and practice is essential for acceptable proficiency but the advantages make this method of insemination more desirable than other known methods. With practice, the skillful technician soon learns to thread the cervix over the catheter with ease. If disposable catheters are used and proper sanitation measures are followed, there is little chance of infection being carried from one cow to another.

Timing of Insemination for Maximum Conception

Since estrus may last from 10 to 25 hours there is considerable latitude in possible time of insemination. Much research work has been conducted on this subject. Controlled investigations were conducted by Trimberger and Davis at Nebraska in 1943. These and other studies show that conception rate is lower when cows are bred prior to midestrus or later than 6 hours after cessation of estrus (standing heat in this case). Maximal conception is obtained when cows are inseminated between midestrus and the end of standing estrus, with good results up to 6 hours after estrus.

Success in insemination timing is dependent upon a good heat detection program. In large herds, this means assigning individual responsibility for heat detection and a continued education program for labor. A successful heat detection program and subsequent proper timing of insemination will pay dividends in increasing reproductive efficiency.

Crossbreding of Dairy Cattle

Under nearly all conditions of climate and management throughout the world, except the very worst, the milk yield of a local cattle breed can be increased by crossing it with an improved dairy or dual-purpose (milkmeat) breed. Thus, as countries with a temperate climate develop their dairy industries they inevitably replace their local breeds by such improved breeds, either by grading up (displacement crossing) or by importing both cows and bulls. The same applies to developed countries in subtropical regions, such as Israel, South Africa, southern Australia and the southern United States.

As long as the European dairy cattle are well fed, protected from disease, shielded from direct sun and given a chance to cool down at night and in the winter, they do not appear to be seriously inconvenienced by high environmental temperatures. It may be difficult to get cows in calf during the hot wet season and there may be a seasonal decline in yield but in general a European dairy breed is the most economic type for these regions. The same applies to those tropical regions in which the climate is moderated by altitude, as in many parts of east and south-central Africa, or by proximity to the sea, as in the islands of the Caribbean and the South Pacific.

However, there comes a point as we approach the deep tropics where purebred European cattle require more elaborate management than is possible under the existing economic and technological conditions. Many trials in east Africa, India, southeast Asia and Central America have shown that crossbreds between local breeds and European dairy breeds not only produce more milk than the local breed but also more than the pure European. Naturally, if the first cross is half way between the two breeds and the backcross to the European breed is half way between halfbred and purebred then the genetic situation is purely additive and grading up to the European breed is indicated. This article is concerned with the case where optimum performance is given by some intermediate between the two breeds, and will discuss breeding schemes which can be used to maintain this hybrid vigour.

Maintaining a Crossbreed Population

There are three ways of mating the F1 females after the initial cross. They can be mated to sires of the local breed or of the improved breed or they can be mated to F1 sires. On an experimental basis, all three matings should be made. This will make possible a comparison between the F1 and F2 and between halfbreds and three-quarterbreds. Hopefully, purebreds will also be present for comparison.

If the halfbred or threequarter is clearly superior then *inter se* breeding within this level of exotic blood should begin at once. There is no point in trying further to refine the exact proportions which may be optimal. There will be so much variation within grades that only an approximation is necessary and selection can at once be based on performance rather than on proportion of blood. Because the well-known Santa Gertrudis beef breed was based on 3/8 zebu and 5/8 British, there has been a tendency to extol this as the ideal proportion and it has been followed in several other new breeds; e.g., the Bonsmara of South Africa and the Jersind and Brown Sind of Allahabad. This is pure formalism. In practice there are three principal ways of maintaining a crossbred population, whether it starts from an F1 or a backcross:

The crosses may be bred *inter se* to produce a new breed which will have, at least initially, either 50 percent or 75 percent of the blood of the exotic breed. The crossbred bulls may be used to mate with, and eventually to grade up, the local cows. The proportion of exotic blood will thus rise gradually to reach the limiting values of 50 percent, or 75 percent when the displacement is complete and a new breed is formed.

The F1 cows can be bred to bulls of the two pure breeds in alternate generations in a crisscrossing system. When equilibrium is reached the cows will alternate in successive generations between 66.6 and 33.3 percent exotic blood.

In systems (1) and (2) 50 percent of the hybrid vigour of the F1 is maintained in the new breed. System (3) retains two thirds of the hybrid vigour. On theoretical grounds this would be the system to use if a large decline in performance from F1 to F2 indicated the presence of hybrid vigour.

However, important criteria are: which system can be operated in practice and which offers the best scope for a selection programme to follow the initial crosses? Effective genetic improvement in milk yield depends firstly on an efficient milk-recording system, and secondly on a bull progeny testing scheme based on the results of this recording.

If such a system is available it can be applied at once to the new breed. The breed formation will have to be made on a scale sufficient to provide opportunity for progency testing several bulls each year. This cannot be restricted to a single small government herd. Within a single experimental herd milk recording is easy enough but cows may be insufficient in number to allow an adequate progeny test of several bulls and a choice between them.

If no milk-recording system is available then it may be better to exploit the improvement programme in another country (or possibly in a more developed part of the same country) by importing improved bulls or their semen and using a crisscrossing or rotational breeding system.

Formation of New Breeds

The activities of breeders and breed societies in western Europe and North America during the last century have taught us to think of breeds as closed population of nearly identical animals. From the present point of view the important characteristic of a breed is that it is self-contained. It generates its own supply of bulls and does not have to rely on an outside source of bulls or semen.

Because of the aura from the past and the emphasis on uniformity there has been a reluctance to take the step of attempting to form a new breed. People are afraid that breeding from crossbreds will lead to excessive variation, and therefore that formation of a new breed is only possible if immense resources of animals are available. In fact, the increase in variation on breeding an F2 generation has been much exaggerated. To begin with, it is only the genetic variation which is increased, and with characters of low heritability this is only a small part of the total variation. Secondly, the obvious effects are due to segregation of colour, pattern and morphological characters (e.g., horns) which are controlled by single genes. The increase in variation between F1 and F2 for such quantitative characters as growth rate and milk yield will be very small, and in view of other sources of variation may not be detectable.

Certainly there is a danger that if the initial crossing is not done on a large enough scale then problems of inbreeding may be encountered. This is particularly the case if the breed is based on a single herd or on only one or two imported bulls. Inbreeding leads to decline in fertility, viability and growth rate to a greater or lesser extent. This has often not been realised and inbreeding has been deliberately employed in order to concentrate the blood of outstanding sires or to produce uniformity in a new breed. But uniformity is an elusive goal, a will-o' -the-wisp. It must not be sought at the expense of productivity. Indeed, it is essential at the outset to have as much variation as possible in order to allow scope for selection. The aim of crossbreeding is to combine the high yield of the European breed with the resistance of the local breed, and intense selection is needed to find the few animals combining both characters.

However, one should not be put off by takes about the formation of the Santa Gertrudis breed, which is said to have been based on a herd of 25 000 cattle on a ranch of 400 000 hectares (a million acres). In fact, only a few hundred of these animals were used for breed formation and the rest of the herd was merely graded up. The lesson from the Santa Gertrudis (as from the Shorthorn) is the danger of concentrating too much on single bulls. Both breeds were linebred to a single bull — *Monkey* for the Santa Gertrudis and *Shakespeare* for the Shorthorn — and it is probably no coincidence that both breeds have at times run into severe infertility problems.

If a breed is to be formed in a single experimental herd it is therefore essential to use a large number of bulls (at least 10-15) in the formative stages. As early as possible the programme should be extended to other herds and bulls should be interchanged between herds. A large cow population will be needed for progeny testing the bulls. Above all, animals should be selected not according to colour, conformation or other fancy points but for important economic characters which define total dairy merit, i.e., milk yield, viability and thrift of the calf, fertility of the cow, ease of milk let-down, udder and teat characteristics (including durability and suitability for machine milking, if appropriate), docility, and possibly also milk composition.

At the present time crossbreeding between zebus and dairy breeds is taking place at several centres in the tropics. It would be most desirable to coordinate the activities within each crossing breed and encourage exchange of breeding stock between countries to ensure a wide genetic base for each of the new breeds.

New Breeds by Introducing European Bulls

There have been several attempts in the past to form new

breeds by introducing European dairy genes into the zebu cattle of India and Africa, but none has come to fruition.

In Tanzania, for instance, Hutchison started a crossbred herd of European dairy cattle x east African zebus in 1946, but it was disbanded in 1966. Earlier the new strain had been officially baptised "Taurindicus". It is clearly easier to name a new breed than to maintain it, and the reports from India contain many attractive and descriptive names such as Jersind, Brown Sind and Karan Swiss which presumably will all be eventually absorbed in the Kamaduk.

In tropical South America one has read of the Ocampo (Venezuela), Jerdi (Brazil), Tropical, Tropicana and Suisbu (Argentina). It would be interesting to have more particulars of these breeds and any others of European x zebu origin. While there are still no good examples of new dairy breeds whose genesis has been planned and executed on the basis of a European male X zebu female cross, there are two interesting examples of populations developed haphazardly from this cross in the past which have remained as distinct entities because of their economic value. These are the Taylor cattle of India and the Hatton cattle of Sri Lanka.

The Taylor cattle of Patna (Bihar state) originate from four Shorthorn and Channel Island bulls introduced by Commissioner Taylor in 1856 for crossing onto the local zebus. According to Sinha (1951) there were in his time about 2,500 Taylor cows, and their milk yield averaged 6–8 kg daily.

The Hatton (or Cape) cattle are said to derive originally from European imports which the Dutch brought to Ceylon via the Cape of Good Hope between 1765 and 1815. Until the more recent imports of European dairy breeds they were the principal milk producers on the island. Neither of these populations has had the benefit of either a scientific breeding programme or a society for breed promotion. That they should have survived so long indicates that they must have considerable intrinsic merit. They deserve to be called "breeds" and merit a suitable improvement plan.

Indeed, a programme of milk recording, progeny testing and breed promotion might even pay quicker dividends than starting now the laborious business of forming a new breed from scratch. Such a programme could, of course, include the introduction of outstanding halfbred bulls in order to benefit from the genes of today's improved dairy breeds.

New Breeds by Introducing Zebu Bulls

The first cattle introduced into the New World (i.e., America and Australia) came from Europe. When it was observed that in the tropical and subtropical areas of these continents they performed less well than in their homelands it was thought that the explanation might be lack of heat tolerance. The remedy suggested was the introduction of zebu blood.

For this purpose the dairy zebus from Pakistan, the Red Sindhi and the Sahiwal, have been the most popular. Red Sindhis were imported into the southern states of the United States for experimental crossing with Jersey, Holstein-Friesian and Brown Swiss. The result of the cooperative experiments carried out in Georgia, Louisiana, Maryland and Texas showed that as the proportion of zebu blood increased, the milk yield fell. It is true that heat tolerance increased, but this did not have any economic advantage. The differences in fertility, if significant, were in favour of the pure European breed.

These experiments have been given up, and the milk yield of European breeds in the North American subtropics is being increased by attention to feeding and management combined with selection for productivity within the hot environment.

Nevertheless, the conclusions of Branton *et al.* include the sentence: "If the hypothesis of lower nutritive requirements is acceptable, then under conditions of adverse climatic conditions and poor nutrition, some proportion of zebu breeding may prove desirable." There are two areas where this has proved to be the case and two new breeds, the Jamaica Hope and the Australian Milking Zebu (AMZ), have been developed.

At the government farm (Hope Agricultural Station) in Jamaica the beneficial effect on milk yield of using halfbred zebu bulls on Jersey cows was recognised before 1920, and in that year a single Sahiwal bull was imported to continue the good work. Fortunately he turned out to be an outstanding animal, and the Jamaica Hope breed owes most of its zebu genes to him. The new breed contains about 20 percent zebu blood and 70–75 percente Jersey, the rest being from Friesian and other breeds. The bulls produced in the government herd were used to grade up the general Jersey-zebu stock on the island and now most of the dairy cows belong to this breed.

Mahadevan gives an excellent short account of the origin of this breed and Mahadevan *et al.* show that its further development is being limited by the restriction of progeny testing of bulls to the government herd. This means that the number of daughters per bull is too low for an accurate assessment, or else that too few bulls can be tested to given an adequate intensity of selection. All farmers should be involved in testing, by using young bulls extensively through artificial insemination and milk-recording their daughters.

In Australia the approach to forming a new breed based on Jersey and zebu has been much more systematic. In fact, the first experimental results were similar to those in the United States — in both cases the F1 had a much lower milk yield than the Jersey. At this point the Americans gave up but the Australians were convinced that zebu blood had something to offer in the warmer, tick-infested parts of northern New South Wales and in Queensland. The breeding system used has been crossbreeding and selection up to the F3 generation, followed by grading up the local population of Jerseys to the halfbred bulls and at the same time progeny testing these bulls.

Feeding and management levels are clearly different in Australia and the United States. In Australia there is more reliance on pasture. The F1 cows also have more difficulty in establishing a lactation in the absence of the calf. In the United States the F2; yielded less than the F1; in Australia the F2 and

F3 yielded more than the F1. This increase was presumably achieved by intense selection among their sires.

New Breeds of Multiple Racial Origin

The new breed formation is concerned with a foundation based primarily on two breeds only — one improved temperate dairy breed and one zebu. In a planned breed formation three or more breeds could be used as a matter of deliberate policy. This will increase the genetic variance and hence the scope for selection, but it will also make the programme more complicated and delay the final birth of the breed. It may also overload the foundation with genes from inferior breeds — note how Hayman had to get rid of the inferior Sindhi crosses.

The Indian Council of Agricultural Research has launched an ambitious programme to create a new breed — the All-India Coordinated Research Project on dairy cattle breeding. It is based on the current crossbreeding between Jersey, Friesian and Brown Swiss on the one hand and various local zebu breeds on the other (Hariana, Sahiwal, Gir, Ongole, Hallikar, Tharparkar) at eight different centres.

In the breeding scheme, the first crosses will be backcrossed to a second European breed and the backcrosses combined in pairs. Thus it will be three generations before the foundation stock is ready. Then will start the important phase of progeny testing bulls on a large enough scale so that those rare animals can be picked out which combine (in some measure at least) the genes for high milk and butterfat yield and growth rate and fertility of the European breeds with those for disease resistance and heat tolerance of the Indian breeds. This is a very ambitious programme. It might be quicker and easier to think in terms of two new breeds — one based on the Jersey for early maturity and butterfat, the other on the Friesian for milk and beef.

Crisscrossing between local and improved breeds

The examples given above show that while it is easy to

increase the milk yield of a zebu breed by crossing with a European dairy breed the difficulty is to maintain the advantage in later generations. It needs an extensive system of milk recording and progeny testing, preferably based on an artificial insemination service so that each bull can have many daughters distributed through several herds. If no such facilities are available it is better to use a crisscross system of breeding.

When equilibrium is reached, the cows in alternate generations have one-third improved or one-third local blood. These two types should be kept in separate herds, the first run with a European bull and the second with a local bull. If this is not possible, the presence or absence of hump (assuming that the local breed is a zebu) will indicate which breed of bull should be used on each cow. If artificial insemination is used the separation into herds is not necessary; the humped cows will be inseminated with European semen and the humpless with zebu.

The total resources of the farm can be used for keeping cows and breeding their replacements. All male calves will be slaughtered and bulls will be brought in as necessary. The continuing genetic improvement stems from these bulls. They will come from herds in which a rigid breedimprovement system is in operation If artificial insemination is available then semen can also be obtained from outside the country, but if the exotic semen comes from a temperate country the question of genotype x environment interaction must be borne in mind — the best bulls as tested under temperate conditions may not be the best in the tropics.

Naturally a source of improved local bulls is also necessary, and this highlights the importance of a selection programme such as that described by Meyn and Wilkins for the Kenya Sahiwal in this issue. Replacement cows will be chosen from the best cows of the two types. The two-thirds exotic may well give more milk than the one-third exotic, but the temptation to choose more two-third than one-third exotics as replacements should be avoided. This would mean a preponderance of onethird exotics in the subsequent generation. It would therefore be desirable to start the crisscrossing by backcrossing half the F1 cows to the local and half to the exotic bull. This would enable the two types to be kept in equilibrium from the start.

Crisscrossing has very definite genetic advantages. By breeding always from crossbred cows it exploits any hybrid vigour in fertility as well as in yield. It can be modified by introducing a third breed into the rotation, e.g., a second European breed, if it appears desirable to maintain a higher proportion of European blood.

However, since the last cross was by the breed with the highest milk yield it is not clear how much of the advantage of the three-breed cross is due to additive gene action and how much to heterotic effects. A comparable system has been in operation among beef-cattle breeders in Africa for many years on an *ad hoc* basis. The defensive, almost guilty way they describe it is entirely unnecessary. The system is genetically sound and should be exploited for both beef and dairy cattle.

References

Amble, V.N. & Jain, J.P., "Comparative performance of different grades of crossbred cows on military farms in India", *Journal of Dairy Science,* 1967.

Anon, "Breeding to improve milk production in the tropics", *Activities in Turrialba,* July 1973.

Bhat, P.N., "Kamaduk: the cow of plenty", *Indian Farming,* October 1972.

Branton., C., McDowell, R.E. & Brown, M.A., "Zebu-European crossbreeding as a basis of dairy cattle improvement in the U.S.A.", *Southern Cooperative Series,* Bull, 1966.

Lasater, L., *The Lasater philosophy of cattle raising,* Texas Western Press, 1972.

Littlewood, R.W., "Crossbreeding for milk", *Indian Journal of Vetenery Science,* 1933.

Madalena, F.E., "Crossbreeding strategies for dairy cattle in Brazil", *World Anim Rev,* 1981.

_______________., "Economic evaluation of breeding objectives for milk and beef production in tropical environments", In *Proceedings of 3rd World Congress on Genetics Applied to Livestock Production,* 1986.

Meyn, K. & Wilkins, J., "Breeding for milk in Kenya, with particular reference to the Sahiwal stud", *Wld Animal Reverence (FAO),* 1974.

Sinha, B.N., "Taylor cows of Patna", *Indian vetenery Journal,* 1951.

Wijeratne, W.V.S., "Crossbreeding Sinhala cattle with Jersey and Friesian in Ceylon", *Animal Production,* 1970.

6

Milking Process

Dairy farmers, with varying levels of skill, knowledge and resources, maximise returns from milk production by influencing lactation through selective breeding and control of reproduction, nutrition, disease and general management.

Methods of Milking

The methods of milking have a particularly important effect because a cow cannot secrete over a period more milk than is removed by milking. Thus, maximising milk removal in ways which are economic will take fullest advantage of secretion potential.

Lactation

Lactation includes both milk secretion and storage in alveolar cells and ducts within the mammary gland, followed by milk ejection (let-down) and milk removal. Milk secretion is continuous and usually at a constant rate for at least 12 hours resulting in a gradual increase in internal udder pressure.

Milk ejection is a neuro-hormonal reflex initiated by various stimuli at milking time. These stimuli, which reflect good husbandry practices, are either natural (inborn) or conditioned (learned by experience), including, for example, feeding and udder preparation. They cause the alveoli and small milk ducts to contract forcing milk towards the udder sinus. Once this has happened most, but not all, of the milk can be removed when external forces such as suckling or milking open the streak canal (teat duct) at the teat end, but at least 10% will be retained in the udder as residual milk.

Milking Intervals

The 10%-20% of the secreted milk which is not expressed from the secretary tissue and is retained in the udder when milking is completed is called 'residual milk' and has a much higher fat content than even the end-of-milking strippings. The quantity of residual milk is proportional to total yield, so that with unequal milking intervals there is a larger net carryover of milk fat from the longer night-time to the shorter daytime interval.

Milking intervals affect the amount of residual milk carryover between milkings. Equal intervals of 12 hours give highest lactation yields but the effect of unequal intervals is small up to 16 and 8 hours and can be minimised if the highest yielders are milked first in the morning and last in the afternoon.

This accounts for the apparent faster secretion rate and higher fat content of afternoon milk production. Milk yields, particularly from higher yielding cows are usually greater when milking intervals are 12 hourly. The effect of uneven intervals is not large up to 16 and 8 hours, and can be minimised by milking the higher yielders first in the morning and last in the afternoon.

Frequency of Milking

As a general rule, herd lactation yields will rise as the frequency of milking is increased. On average, the rise in milk yields will be between 10% and 15%, the largest increases occuring amongst heifers. The chemical composition of the milk (fat and solids-not-fat) will be unaffected. Recent commercial data from developed dairying areas also reveal that, on average, up to 10% increase in yield is required to cover the extra costs of milking three times daily. The full benefit of the increased frequency is obtained by milking three times daily throughout lactation rather than reverting to twice daily when milk yields begin to fall. The reasons for the increase in lactation yields are inconclusive; the most likely being the more frequent removal of secretion inhibiting substances which begin the drying-off process.

Incomplete milking

There are two forms of incomplete milking. One is that excessive amounts of residual milk are retained in the udder because of inadequate milk ejection stimuli or the inhibitory effects of adrenalin secreted by cows becoming frightened and upset during milking, or even by slow milk removal. The other form is that some of the available milk is left in the udder when milking ceases, ie., the so-called 'strippings'. The modern milking machine is designed to remove 95% of available milk without recourse to additional cluster weight or manual assistance. Hand stripping, particularly with the finger and thumb should be avoided. The amounts of strippings are likely to be small even in relation to normal levels of residual milk and if not removed are unlikely to affect significantly either the lactation yield or quality of milk.

Milking Routines

The aim of an efficient and effective milking routine is to leave the least amount of residual milk in the udder. This, in itself, is a measure of good stockmanship. Milk ejection can be stimulated manually by a series of activities carried out by the person doing the milking. The amount of residual milk is inversely proportional to the strength of the conditioned stimuli signals, which are developed into a regular, repetitive milking routine, including such activities as feeding and udder preparation. The stimulation response is transitory, the maximum effect declining within a few minutes of milk ejection occuring. Therefore delayed milking will reduce the amount of milk removed. The internal pressure of milk within the udder peaks between one and two minutes after milk ejection and therefore milking should be completed as soon as possible after this occurs.

Cows are creatures of habit and consequently changes to the routine should be made gently and quietly. It is important to avoid any circumstances which upset or frighten them causing the release of adrenalin which adversely affects the circulatory and musculatory systems, thus restricting effective milk ejection and prolonging the duration of milking. The response of cows and those milking them to a pleasant and stress free environment will be measured in terms of production levels.

Because residual milk and strippings have fat percentages that normally exceed 10%, incomplete or slow milking can reduce markedly the fat content of the milk at any particular milking. However, it is important to recognise that milk fat retained or left in the udder is not lost but will be obtained at succeeding milkings. In fact, although management factors (eg. varying milking intervals and milking frequency) may alter the fat content of milk at one milking, the average fat content over a period of time will be unaffected. On average, the fat content of milk obtained must be the same as that secreted into the udder. The concentrations of protein, lactose and other solids-not-fat are unaffected by changes in milking management either at one or more milkings.

Methods of Machine Milking

Hand milking should always be done using clean, dry hands. Preferably, milk with the full hand and avoid end-of-milking stripping with the finger and thumb. Rear quarters should be milked first as they contain most milk and the milking bucket hooded to reduce contamination from dust and udder hairs.

Methods of machine milking are designed to create a pleasant milking sensation for the cows and to avoid any possible hazard to udder health. It is most important that milking is done with a well designed, carefully cleaned and properly maintained machine which is operated strictly according to the manufacturer's instructions.

A skilled operator pays particular attention to careful cluster attachment and removal from the udder. During cluster attachment it is essential to ensure that the vacuum cut-off arrangements to the clawpiece are effective so that excessive volumes of air do not enter and cause vacuum fluctuations in the main vacuum pipeline system, as this could increase mastitis incidence. Attach each teatcup carefully starting with the two furthest from the operator. The clusters are removed as soon as milk flow ceases, avoiding excessive air entry through the teatcups by cutting off the vacuum supply before gently but firmly pulling the teatcups from the udder. During milking, any teatcups which slip from the teats should be readjusted immediately and any clusters which fall to the floor should be cleaned and re-attached without delay.

Milking Machines and Equipments

The basic layout of the three main types of milking machines are the same. Each has a pump to remove air from the vacuum pipeline, a vacuum regulator and a container to collect the milk that comes into the teatcup assembly during milking.

Basic construction

The principle of machine milking is to extract milk from the cow by vacuum. The machines are designed to apply a constant vacuum to the end of the teat to suck the milk out and convey it to a suitable container, and to give a periodic squeeze applied externally to the whole of the teat to maintain blood circulation.

A milking machine installation consists of a pipework system linking various vessels and other components which together provide the flow paths for air and milk. The forces necessary to move air and milk through the system arise from the fact that it is maintained at a vacuum. Thus it is atmospheric pressure which forces air, and intra-mammary milk pressure which forces milk, into the system and the combination of these forces causes flow. To be a continuous operation it is necessary to remove air and milk from the system at appropriate rates.

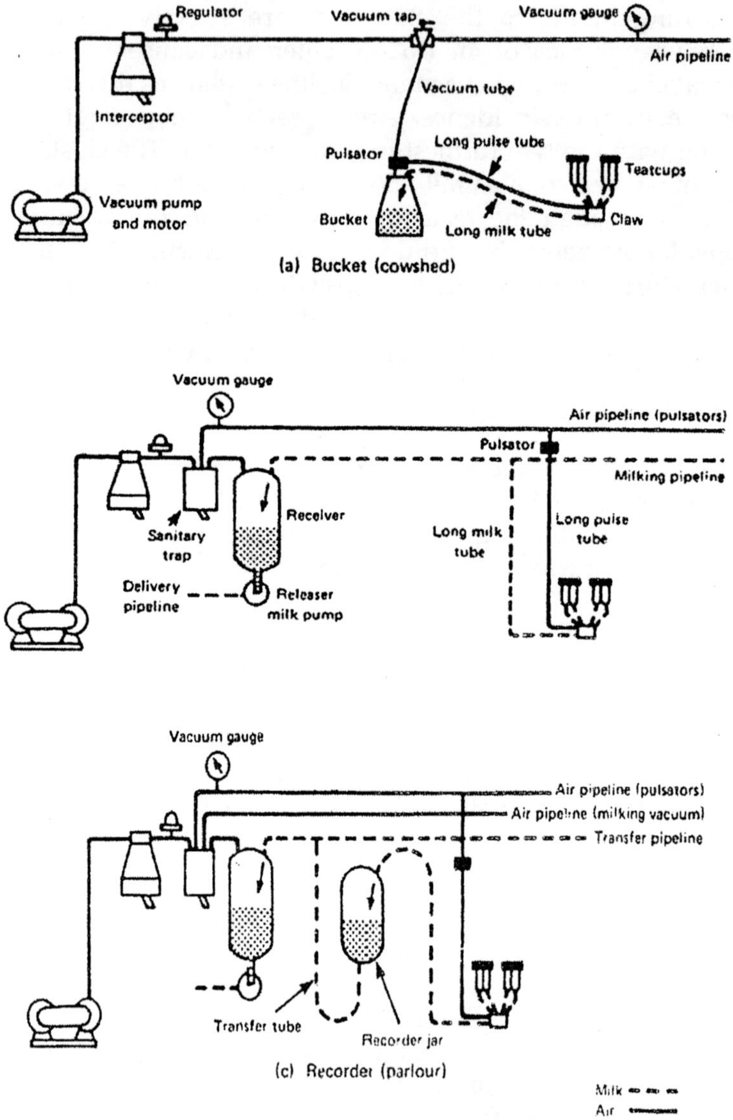

Figure 1. Principal types of milking machines

Although milking machines have now developed into systems that show considerable diversity they have the same basic components. The air is removed by a vacuum pump at a constant rate. In a bucket or direct-to-can machine milk is removed from the system by disconnecting the milk container; in milking pipeline and recorder machines the milk is removed by a milk pump or releaser.

In the bucket (or direct-to-can) machine the milk enters the teatcups and travels through the short milk tubes to the claw where air is admitted and the milk and air travel along the long milk tube to the bucket (or can). The milk remains in the bucket (or can) and the air separates to pass up the vacuum tube to the vacuum pipeline. The pulsator which is usually fixed on the bucket lid admits air intermittently and this passes along the long pulse tube to the teatcup chambers. To control the vacuum at a predetermined level air is also admitted to the system through a vacuum regulator which is fitted on the vacuum pipeline near to the milking points.

Vacuum and Milk Flow

When the milk from the claw is raised to a pipeline this can markedly reduce the vacuum at the teat because of the weight of milk in the long milk tube. The reduction in vacuum can be much reduced by bleeding air through a small hole in the clawpiece

In addition to the designed sources of air admission, air can be drawn into the teatcups past the teat and also when a milk container is changed or emptied. In a poorly maintained machine there may also be inward leakage of air at joints or points of damage. To maintain the working vacuum the vacuum pump extracts the air admitted into the system by compressing it so that it can be discharged to atmosphere.

In pipeline milking machines the flow pattern is similar to the bucket machine except that milk and air from each claw flow either directly to a recorder vessel where air and milk are separated, and/or through the milking pipeline to a common receiver vessel where milk and air are separated. There is no further air admission at this point when a motor driven releaser milk pump is used to empty the receiver. Other types of releaser (eg. pulsator controlled spit chamber and double chambered weight operated) admit air.

Where air and milk are transported together the flow pattern becomes complex depending on various factors particularly the volume of air relative to milk or air:milk ratio. Air is normally admitted in to the claw at a rate of 4 to 8 l/min. A milk flowrate for a fast milking cow will be about 6 l/min, giving an air:milk ratio of 0.7:1 to 1.2:1. Towards the end of milking when the milk flowrate has decreased to 0.25 l/min the ratio becomes 16:1 to 32:1.

The air:milk ratio becomes important where milk has to be elevated from the claw as in milk pipeline and recorder machines other than those with low level milk pipelines. Elevating a liquid, as distinct from a gas, involves a loss of potential energy and this is compensated for by a change of vacuum. Thus elevating a column of milk in a vacuum system through 1 m height reduces the vacuum by about 10 kpa. Therefore if the vacuum at the top of the column is 51 kpa it will be only 41 kpa at the bottom. This vacuum drop is markedly reduced by the admixture of air. If the air:milk ratio is 1:1 the weight of milk in the column is halved and the vacuum drop becomes only 5 kpa; if it is 9:1 the vacuum drop is only 1 kpa.

Under vacuum liquids cannot flow against gravity (ie. uphill) except as a column which fills the bore of the tube. Where the tube contains air and milk the liquid forms plugs which are separated by pockets of air in the proportion determined by the air:milk ratio.

Vacuum measurement

Vacuum is a pressure below atmospheric pressure. It can be measured as pressure difference with a mercury manometer in mm Hg. The standard pressure is now kilopascals (kPa) with 100 kPa equal to the pressure difference between atmospheric pressure and absolute vacuum.

Vacuum is a pressure below atmospheric pressure, the term "negative pressure" is sometimes used but in milking machine terms it may be considered to mean "vacuum" measured on a scale in which atmospheric pressure at the time and place of measurement is zero vacuum.

Vacuum can be measured in a variety of units. A commonly used measure is the linear difference in height between two columns of mercury in a 'U' tube when one of the columns of mercury is subjected to a vacuum and the other open to atmosphere. The difference in height of the levels is supported by atmospheric pressure.

In the past the most commonly used units have been inches, millimetres or centimetres of mercury (in Hg, mmHg or cmHg). Units now adopted by the International Standards Organisation (ISO) for International Standards of milking machines for vacuum measurement are kilopascals (kPa) with zero (0) kPa being equal to atmospheric pressure and 100 kPa absolute vacuum. Equivalent relationships for values of vacuum levels are:-

$$1 \text{ mmHg} = 0.133 \text{ kPa}$$

$$1 \text{ inHg} = 3.386 \text{ kPa}$$

Equivalents for vacuum levels of 50 kPa and 44 kPa that are the most commonly used levels for milking cows:

$$50 \text{ kPa} = 14.8 \text{ inHg} = 375 \text{ mmHg}$$

$$44 \text{ kPa} = 13.0 \text{ inHg} = 330 \text{ mmHg}$$

Action of the Milking Machine in each Pulsation Cycle

Pulsator connects pulsation chamber to vacuum, liner opens and milk flows.

Pulsator connects pulsation chamber to atmosphere, liner collapses, squeezes the teat duct and prevents milk flow.

Milking Rate

Milk flow from the teats increase with:

— increasing vacuum but strippings also increase. (Normal vacuum range 40-50 kPa).
— increasing pulsation rate but this increases air to be pumped from the machine. (Normal rates 50-60 cycles/ min)
— widening pulsation ratio, ie liner open to liner collapse time. (Normal range 50/50 to 70/30)

Teatcup Liners

Teatcup liners have important effects on milk flow and completion of milking.

— the liner mouthpiece affects the quantity of strippings
— narrow bore (<24 mm) and low tension (stretch) milk more slowly.
— choose liners appropriate for the size of cows teats to be milked.

Action of the Milking Machine

The principles of machine milking were established many years ago and the basic method described below, is used in virtually all commercial milking machines although in a minority some modifications are made. The teatcup liner is the only equipment that comes into contact with the cows teats. The continuous vacuum within the liner causes the teat duct (streak canal) to open and the milk to flow because of the pressure difference between the milk in the teat and vacuum. To prevent damage or pain to the teat that would be caused by the continuous vacuum a system called pulsation is used. This makes the liners collapse on and below the teats about once each second massaging the teat and maintaining a more normal blood flow. In each pulsation cycle milk does not flow from the teat when the collapsed liner squeezes the teat duct.

Providing the cows 'let down' (ejection) has occurred the flow rate from the teat depends largely on the bore of the teat duct which is an inherent factor and not subject to management practices or training. Flow rates are also influenced by the mechanical properties of the milking machine. After the teat cups have been attached the flow rate reaches a maximum in about one minute, usually within the range of 2-5 kg/minute and the total milk flow period will range from 2 to about 8 minutes depending upon milk yield.

Flow rates decline at the end of milking and when flow ceases there is usually a small amount of milk trapped in the sinus of the udder which can be removed by pulling ownwards on the clawpiece and massaging the udder (ie. machine stripping). With modern designs of liner the quantity of strippings is small (ie. less than 0.3 kg) and machine stripping is not usually practiced. The small amounts of milk that are left do not affect milk yield or the average chemical composition of the milk obtained or mastitis.

Vacuum Fluctuations

The fluctuations in vacuum in the teatcup liner have important effects on mastitis and milk flow. There are two types.

Irregular fluctuations

These occur when the teatcup liners slip or fall from the teats or air enters when milking units are changed carelessly. Vacuum recovery is slow if there is inadequate vacuum pump capacity.

Cyclic (regular) fluctuations

The cyclic movements of the liner in each pulsation cycle increase and decrease the volume of the liner under the teat. When milk is flowing this can cause marked changes in vacuum below the teat. This can be reduced by:

- — using wide bore short milk tubes (>8 mm)
- — ensuring claw or short milk tube air bleeds are not blocked.

The main milking machine factor affecting milk flow rate is the liner vacuum. Raising vacuum levels gives faster milking but also increases strip yields and in practice a compromise level of about half atmospheric pressure is used (ie. 40-50 kPa,

300-375 mm/Hg). The pulsation characteristics also affect flow.

An increased pulsation frequency (rate) gives faster milking but because this greatly increases the air admission in the machine and therefore the required pump capacity it is usual to keep pulsation rates at 50-60 cycles of liner opening and closing per minute. Because milk flow ceases in each pulsation cycle when the liner is collapsed on the teat, faster flow rates are obtained by using a wider pulsation ratio (ie. ratio of liner open time to liner collapsed time).

For udder health reasons the ratios are usually not greater than 70:30. The design of the liner can also affect the flow rate but modern liners tend to have similar flow properties. Narrow bore (<24 mm) liners and those with low tension in the barrel (ie. not stretched in the teat cup) milk more slowly. The most important characteristic of the performance of a liner is the amount of strippings left at the end of milking which is mainly determined by the dimensions and hardness of the mouthpiece. Liner design is largely empirical and farmers determine the best liners for minimum strippings by trial and error.

Although the teat cup liners are connected to a pipeline maintained at a constant vacuum level there can be considerable vacuum fluctuations in the liners, mainly due to the movements of the liner wall brought about by pulsation. When the liner is opening and the milk is moving away from the teat along the short milk tubes the vacuum below the teat will increase markedly due to the increased volume of the liner and the kinetic energy of the milk in transit. These variations in vacuum occur with each pulsation cycle and are called 'cyclic fluctuations'. The fluctuations are increased with adventitious air admission that occurs when liners slip on the teats or when machines are removed from adjacent cows udders (irregular fluctuations). These cyclic and irregular fluctuations generate impact forces which are important factors causing mastitis.

Various methods have been used to reduce fluctuations to prevent them having deleterious effects. The most important are the provision of adequate airbleed holes in the claw or short milk tubes to aid milk flow and prevent flooding in the liner. Also useful has been an increase in the internal diameter of short milk tubes (eg. greater than 8 mm). These can be reduced by other modifications. In some designs the basic system of machine milking has also been modified to give lower levels of vacuum at the start and end of milking when there is no milk flow or by the inclusion of a positive pressure phase in each pulsation cycle to give increased let down stimulation.

INTERCEPTOR

Components of Interceptor

Vacuum System

The function of the vacuum pump is to extract air from the pipeline system and in the majority of milking machines it is a rotary exhauster driven from either an electric motor or a stationary internal combustion engine, using pulleys and V-belts. This gives flexibility with speeds ranging from about 800-1500 rev/min and a corresponding ange of pump capacities to suit several sizes of milking machines. A few vacuum pumps are direct shaft coupled; their speed is then motor speed and cannot be changed. When operating with a vacuum at the pump inlet suitable for milking cows (50 kPa), the pumping capacity should be sufficient for the total number of milking units connected.

This will range from 70 litres of air per minute per unit for installations with 20 units to 85 per litres per minute for the smaller plants with 5 units. It will vary according to individual manufacturers design of equipment and the additional ancillary components included. Oil lubricated high speed vacuum pumps operate at speeds of between 850- 1440 revs/min. For each 0.75 kW (1 h.p.) of rated electric motor requirement the extraction capacity is approximately 280 litres of air/min. The pump should be fitted with drive belt guards, an effective silencer and, for oil lubricated pumps, an oil separator.

Also, a nonreturn valve should be fitted in the exhaust pipe to prevent reverse rotation of the vacuum pump when switching off at the end of milking. This is to avoid debris from the silencer/exhaust pipe being sucked back into the pump. To prevent solid and liquid material being drawn into the pump from the milking installation, an interceptor/trap vessel with a capacity of not less than 15 litres is fitted in the main vacuum line adjacent to the pump, with provision for draining and cleaning. It should also have a float valve to shut off the vacuum in the event of it filling with liquid.

Vacuum pump

Vacuum regulator is an automatic valve fitted into the main vacuum pipeline near to the milking units that is designed to maintain the working vacuum level in the milking machine, despite the varying air usage during milking. It operates by admitting air into the pipeline when the vacuum increases above the predetermined level. The valve is held closed by a weight or spring until the vacuum in the system overcomes the closing force, allowing atmospheric pressure to open the valve. Weight operated regulators have a maximum capacity of approximately 550 to 700 litres/min.

Spring operated regulators have capacities of up to 1400 litres/min. Servo or power operated regulators, which usually have remote sensing, utilise a small intermediate pilot valve to actuate the main air inlet valve. These regulators sense pressure changes in the vacuum pipeline at a position where more stable conditions might be expected to exist. They are readily adjustable with capacities varying from 1000 to 5000 litres of air/min. A good regulator should be able to control the vacuum to within plus or minus 2 kPa of the working vacuum, be stable and not leak by more than 35 l/min, when nominally closed. All vacuum regulators have an air filter to filter the incoming air. The vacuum gauge indicates the vacuum in the pipeline system. Its most important function is to indicate abnormal levels and fluctuations in vacuum, eg. serious air leaks, dirty regulator and slipping vacuum pump drive belts.

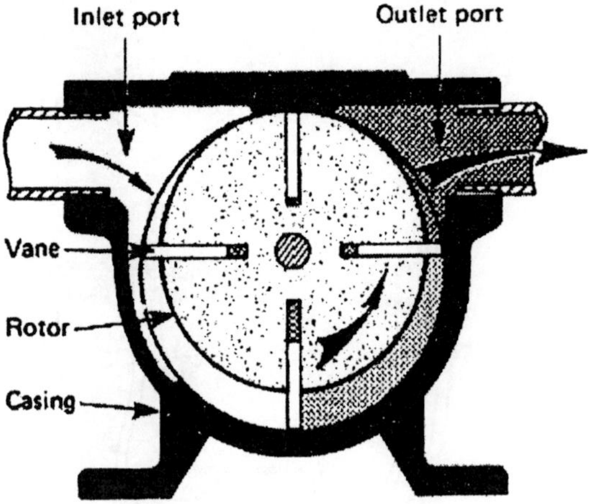

Figure 2. Vacuum pump

Universally the Bourdon type gauge is used. It should not be less than 75 mm in diameter, graduated at intervals of 2 kPa and be adjustable. Most gauges have dual graduations, kilopascals (kPa) and millimetres of mercury (mmHg), and have lines on the dial indicating the working vacuum level.

PULSATORS

Pulsators are valves that cause the liners to open and close on the teat once each second (ie. pulsation) by connecting the pulsation chamber of the teatcup to vacuum or atmosphere.

Pulsation Chamber Vacuum

The pulsator is a simple valve that alternately admits air and vacuum into the pulsation chamber formed between the rubber liner and he shell. This causes the liner to open and close during milking. The valve is activated either by a pneumatic (vacuum) or an electrical signal from a pulsator controller to give a frequency of 40-50 cycles per minute and a ratio (open to closed) of 1:1 to 2:1 (sometimes called 50:50 and 66:33). Relay pulsators are activated from a central pulsator controller, and self contained pulsators have built-in controllers.

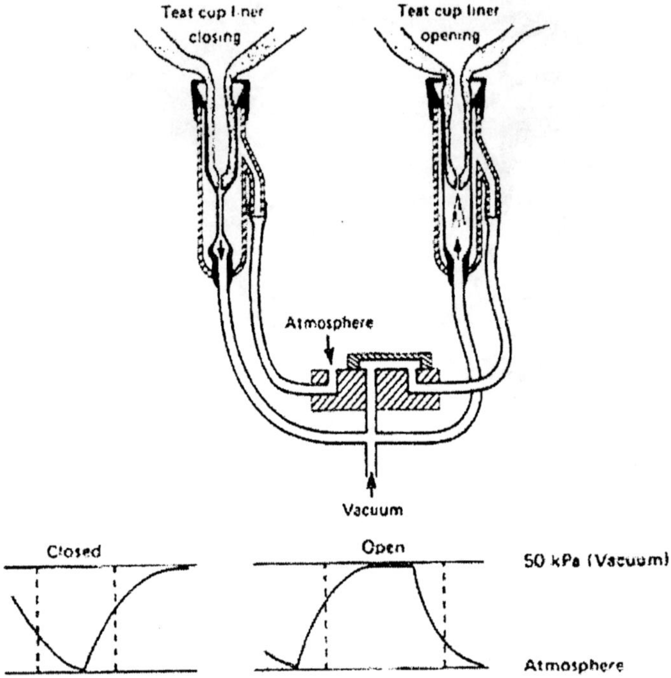

Figure 3. Pulsation chamber vacuum

Most relay pulsators, electrical and pneumatic, pulsate the four liners of a cluster together (simultaneous pulsation). Self contained pulsators are better adapted to pulsating liners in pairs with two open and two closed (alternate pulsation). In the slide valve mechanism of self contained pulsators, the slide valve is arranged to produce a predetermined ratio. With a higher vacuum under the teat than in the pulsation chamber (sometimes due to milk flowing downhill) the liner will not open fully, slightly pinching the end of the teat, resulting in possible teatend damage.

Conversely elevating the milk can reduce the vacuum under the teat, with the result that the liner does not fully collapse. This will influence the designed ratio of open to closed of the liner. These adverse conditions are minimised by ensuring that there is an air admission hole into the bowl of the claw to admit air within the range of 4-10 litres/min as recommended in milking machine standards. The pulsation rate of pneumatic pulsators is determined by a restriction in the air or oil damping system of the pulsator.

With all pulsators, the size of the valve ports, length and bore of the pulse tubes determines the rate of extracting and admitting air. This determines whether a pulsator can operate more than one cluster and is a function of the basic design. There are however solid state electronic pulsators that control the rate and ratio which provide alternate pulsation and can operate two milking units. Air filters are fitted in the air intake to prevent dust getting into the alve mechanisms which should not be lubricated unless recommended in manufacturer's instructions.

Milking Cluster

Consists of four teatcup assemblies each having a rubber liner and connected to vacuum by rubber tubes and claw. The air admission hole to stabilise the vacuum must be kept clear.

The cluster which attaches to the cow, consists of four teatcup assemblies (each having a shell, a rubber liner and a short milk and short pulse tube), a claw, a long milk tube and long pulse tube(s). Teatcup shells are normally made of stainless steel.

Plastics or a combination of plastics and metal are also used. The liner is a flexible rubber sleeve having a mouthpiece, and when assembled in the shell under tension, forms an annular space (pulsation chamber) between the liner and shell. This pulsation chamber is connected to the pulsator through a nipple on the side of the shell via the claw. The teatcup assemblies are connected by short milk and short pulse tubes to the claw, which is connected to the milking and pulsation vacuum by a long milk tube and long pulse tube(s).

To stabilise the vacuum in the teatcups during milking, the claw has a small air admission hole, about 0.8 mm in diameter, which admits approximately 7-8 litres of air/min into the bowl of the claw. This air helps to carry the milk away, preventing flooding and violent vacuum fluctuations.

The claw is made of stainless steel or a combination of plastics and stainless steel, and usually weighs about 0.5 kg and the total all up weight of a milking cluster is about 2.5 kg. The weight of a milking cluster is important and the correct weight relates to the design of liners. Too little weight gives incomplete milking because of high levels of strippings, too much weight will result in milking units falling off during milking.

The bore of the rubber short milk tubes should not be less than 8 mm and the short pulse tubes not less than 5 mm, and the long milk tube should not be less than 12.5 mm. The effective claw bowl volume should not be less than 80 ml.

Milk and Air Separation

In pipeline milking installations it is necessary to include a method of extracting milk from the vacuum system. A receiver vessel is fitted to act as a milk reservoir and air separator and from this the milk is pumped out. It is made of either glass or stainless steel and may have a capacity of 35 to 160 litres or more depending on the method of cooling and storing the milk.

During milking the weight or level of milk in the receiver is used to start the milk pump (releaser). Releaser milk pumps for extracting milk out of the vacuum system are: centrifugal, with capacities of at least 4550 litres per hour, or diaphragm, with capacities of approximately 2000 litres/hr for a single ended pump. Other methods for extracting milk from a vacuum system are double chambered releasers with weight operated mechanical valves and the 'spit chamber' releaser.

These do not require electricity and the latter utilises a pulsating vacuum to alternately open and close flap valves allowing the milk to drain out of the second chamber.

In addition to the interceptor for the vacuum pump, a sanitary trap, is fitted in the vacuum pipeline adjacent to the receiver. This is a glass vessel of not less than 3 litres capacity that separates the part of the milking machine through which milk passes from the air system, preventing movement of liquid from one to the other. If milk enters the sanitary trap it is an indication of a fault in the machine. Therefore, it should be mounted within sight of the milker and be fitted with a float ball to shut off the vacuum. The vacuum connection between the sanitary trap and the receiver should slope away from the receiver to the sanitary trap.

Recording and Sampling

In pipeline recorder milking installations, milk flows directly from the cluster into a rigidly mounted calibrated glass recorder jar where the milk is intercepted and held to allow measurement of the total individual cows milk yield. This is done by using either the calibrations etched on the side of the glass jar or the weight of the contents for jars that are supported on electrical strain beams.

In direct to pipeline milking machines not fitted with recorder jars, milk yields are measured with milk meters. There are several designs of meters. Basically there are two distinct methods used. (1) Proportional flow; these collect a constant proportion (about 2.5%) of the milk flow into a calibrated flask the contents of which are recorded manually or electrically. (2) Batch; the total flow of milk is continuously measured in discrete batches and the aggregate of the batches recorded.

Proportional flow meters used for manual recording are relatively simple and inexpensive. Electrical recording of milk yields, depending on complexity can be costly and are usually linked with electronic management data recording systems.

All these methods of milk recording have the facility for collecting a milk sample for chemical analysis and are required to be within set limits of accuracy for yield and ability to collect a true representative sample of the milk. These limits are laid down in an International Standard for milk yield recording equipment by I.C.R.P.M.A.

Milk Flow Detection and Cluster Removal

With all milking units, except those using transparent recorder jars, there is some provision made for observing milk flow. This can be a metal claw fitted with "windows", or a claw made in a transparent material or, a short length of clear tubing fitted in the long milk tube. Observing flow in these ways can be misleading and there are now milk flow indicator devices that work on the basis of milk flowing into a small container with a submerged metering orifice for maintaining a milk level. Variations in this level can be observed or detected electrically, using a magnetic float switch or electric probes.

In modern large milking installations, milk flow detectors are used to sense the end of milk flow (less than 250 g/min) and activate automatic cluster removal (A.C.R.) equipment. This equipment prevents overmilking and enables the operator to use more units by reducing the workload. It eliminates waiting to observe when an animal has finished milking and manually detaching the cluster.

Electrically operated milk yield recording equipment eg. glass recorder jars on strain beams and milk meters can be used as flow detectors and linked to A.C.R. equipment. In this way the milking cluster is removed when flow rate has fallen below a predetermined level. A useful device that can be fitted in the long milk tube is a small in line filter used as a mastitis detector. It retains milk clots enabling clinical mastitis to be detected and also acts as a coarse filter for removing extraneous material which is an advantage when using flow detectors or milk meters that have small metering orifices etc.

The filter screen is easily removed for rinsing off the clots after each cow and the detector is cleaned in situ by the normal cleaning system of the milking machine.

To reduce or eliminate the effect of vacuum fluctuations that occur during milking "shields" can be fitted in the bottom of the liners near the outlet preventing contaminated milk returning up the short milk tube from impinging on the teat end. The shield consists of a disc of stainless steel or plastics of a diameter that leaves a space etween the edge of the disc and the liner wall not less than the internal diameter of the milk outlet.

A more positive method to prevent the reverse flow in a conventional cluster is to fit non-return valves in the claw or short milk tubes together with a small air admission hole between the liner and the valve to maintain normal milk flow. A system of milking with non-return valves but without the air admission holes has been investigated and preliminary results show the development could have advantages.

Water Heaters

The quantity of hot water required depends on the size of milking machine and the cleaning methods used. It is essential to have an adequate supply. The methods of heating can be, a solid fuel or oil fired boiler, an open top un-insulated electrical wash boiler or a self contained insulated electric water heater. Complete with time switch, thermostat and automatic water filling.

The electrical heating element should be fitted with a clear space beneath it for the hard water scale that will collect.

Water Heating

For nearly all milking machine cleaning methods an adequate supply of hot water is necessary. Water heaters vary from simple free standing insulated or uninsulated tanks filled with a hose and manual control to insulated tanks automatically filled and heated at predetermined times.

The water capacity required is dependant on the washing system used (hand or in-place cleaning) and the number of units to be washed. For in-place cleaning the amount of water required for each milking unit will be 12 literes of water at 85°C for recirculation cleaning and 18 litres at 96°C for a single pass "boiling water" cleaning system. Additional hot/warm water may be required for udder washing and calf feeding. Whatever cleaning methods are used the water heater should be capable of producing near boiling water for the purpose of giving the installation a periodic heat treatment. The temperature of the wash solution in closed pipe work is affected by the vapour pressure of the liquid.

Water at normal atmospheric pressure at sea level boils at 100°C (212°F) and at a vacuum of 50 kPa (14.8 inHg) it boils at 81.7°C (179°F). Thus when cleaning pipeline milking machines with water solutions under vacuum, the maximum temperature which can be obtained ill depend on the operating vacuum level in the installation, ie., the higher the vacuum level the lower will be the temperature that can be obtained. For similar reasons, at an altitude of a 1,000 metres where atmospheric pressure is reduced to about 89 kPa (12.9 lb/in2), water in an open vessel will boil at the lower temperature of 96°C (205°F) instead of 100°C (212°F) at sea level.

Although any fuel can be used for water heating the most commonly used is electricity for cleanliness and convenience. In areas with hard water supplies, a hard scale will build up in the tank and on the heating element. The electrical heating element should be fitted with a clear space beneath it (approx. 10 cm) and be of a type that will flex during heating and cooling, breaking off the hard water deposits that will fall into the space below. Provision should be made for easy removal of this accumulated scale.

For economy any water heater should be well insulated, fitted with a thermostat and preferably a time switch. In soft water areas tanks that are manufactured in galvanized sheet metal are not suitable and copper should be used.

Maintenance

The basic layout and operation of milking machines is normally straight-forward and similar for all standard milking machines. However, it is important that a correct maintenance routine is followed. Faulty milking machines can result in poor milk let down, slow milking, milk of high bacterial count and mastitis. Maintenance is a responsibility of the person regularly using the equipment with simple checks made at each milking and more detailed ones at weekly or less frequent intervals.

The maintenance instructions described deal with the essential routines and do not include checks on the many optional items not fitted in large modern parlours, (eg. automatic cluster removers, milk meters). These vary considerably and the maintenance procedure should be obtained from the supplier. Because the overall performance can be assessed only with special measuring equipment a fullscale test should be carried out by the manufacturer, installer or extension service at least once per year.

The following routines should be read in conjunction with maintenance procedures provided by the suppliers of the milking equipment.

At Each Milking

Confirm that wash water is not retained in milking machine pipelines, particularly the milk lines. This water is likely to have a high bacterial count and reduce the quality of the milk. If water is found drain and flush the milk lines with hypochlorite solution (100 ppm) before milking. In most countries it is an offence to add water to milk and residual water in a milking machine may be sufficient to fail the freezing point of milk test. With some machines and washing systems water can enter the pulsation chambers of the teatcup, damage pulsators, affect the pulsation of the liner and gain entry into the main vacuum systems giving high levels of bacterial contamination. Check the clusters for split liners.

Before starting the vacuum pump confirm that the interceptor is empty either by opening the drain valve or detach the trap for inspection.

Confirm that the oil level in the vacuum pump oiler is at a satisfactory level, ie., not less than half the capacity of the oil reservoir.

Observe the vacuum gauge, the working level should be reached within five seconds, or consistant with the time stated by the manufacturer when the equipment was installed. If there is a delay bove that expected, check for air leaks at open stall or drain cocks, a misplaced receiver or bucket lids and where a centrifugal milk pump is used, for failure of the non-return flap valve to close. If all these items are in order check that the vacuum pump drive belts are not slipping. The working vacuum should be constant and the reading on the vacuum gauge should be the same before and after putting the milking clusters on the cows. Faults found should be corrected at the earliest opportunity.

When all units are in operation and the working vacuum is correct, listen to the vacuum regulator. If there is no sound of air entering the regulator, a serious shortage of pump capacity is indicated. If on starting the vacuum pump the vacuum level rises first to a high level then drops to the working vacuum level the regulator is sticking and requires cleaning.

Check operation of the pulsators. The sound from the pulsators should be regular. When noise is abnormal look for damaged/split rubber pulse tubes particularly at the nipples on the claw, or faulty pulsators. If there is any abnormality check pulsation of liners with thumbs. Examine the milking cluster for damaged rubberware, particularly the short milk and air tubes at the claw nipples.

Condensation or moisture from milking or plant cleaning can collect in the vacuum pump. This rusts the pump body producing, a condition which reduces performance and ultimately would require pump eplacement. This can be avoided by allowing the pump to run for approximately 10 minutes after milking to exhaust the moisture and dry out.

After switching off the vacuum pump examine the interceptor. If milk is found in the interceptor, this may indicate a split liner or malfunctioning of the milk pump. Find the cause, wash out vacuum line and interceptor.

Weekly

To test the pulsation and liner wall movement use the thumbs to sense the movement of the teatcup liner walls. This is done by inserting the thumbs into diagonal pairs of teatcups in turn with the others hanging in the cut-off position; the controls being set for milking. To make certain that the pair of teatcups not being checked are sealed they can be closed with rubber plugs (bungs). With experience abnormal action of the liner is readily detected. If the liners open and close completely and the action of the machine is not painful it will normally be satisfactory for milking. Failure to collapse indicates that the air inlet to the pulsator is blocked or that the pulsator valve is not working properly and the pulsator must be replaced. If this is discovered the operator will find that the cows are uncomfortable and restless and kick clusters off.

The ratio of the liner open to closed cannot be confirmed by this test but it should be possible to detect that the collapse or closed phase is shorter than the open, milking phase. If the pulsation appears to be unsatisfactory check for damaged short and long air tubes.

When checking the liner wall movement confirm that the speed of pulsation is correct. Slow pulsation will give longer milking times. In self-contained individual pnuematic pulsators this slow action is usually due to a low vacuum level or a dirty slide valve. Some designs of pulsators have a speed adjustment facility, this should be reset if incorrect. Most designs of pulsators are designed to operate in the range of 50-60 pulsations per minute. The pulsation speed can also be checked during milking by a slight pressure on a short pulse tube with thumb and fore finger.

Vacuum Pump Drive Belt and Oil Level of Pump

Check the Vee drive belt tension on vacuum pumps. Deflect the belt downwards at a point midway between the pulleys. The movement or deflection possible from the rest position should not exceed 15 mm (0.6 in). A slack drive belt will slip, reducing the vacuum pump performance. If serious belt slip is occuring there will be visible particles worn from the belts on the floor. If adjustment is needed care should be taken to keep the pulleys in alignment. Check the oil level in oil reservoir and if necessary refill.

Filters

Air filters on pulsators and vacuum regulator must be kept clean. The frequency of cleaning depends on whether they are operating in a clean or dusty environment. Most modern components are fitted with foam plastics air filters or screens. Brushing or blowing will remove light dust deposits. Heavily contaminated foam plastics filters should either be renewed or washed in a warm detergent solution and thoroughly dried before re-assembling into the pulsator.

Monthly

Dismantle weight operated vacuum regulators, clean valve face and valve seat with a solvent (methylated spirit). Care must be taken to avoid amage to valve surfaces. Remove difficult deposits with metal polish. Clean regulator filter before the regulator is reassembled and ensure that the weight is evenly balanced.

Servo or power operated remote sensing designs of regulators should be serviced according to the manufacturers instructions as correct reassembly is not always easy. They will require the conical valve face to be cleaned and rubber diaphragm(s), if applicable, inspected. DO NOT oil valve or valve guides.

Check that damping rings or rubbers etc are not damaged. Particular care must be taken when reassembling Servo

Regulators. Observe closely during dismantling how rubber diaphragms, pilot valves etc, are fitted, so that they can be assembled correctly.

The vacuum line and vacuum pump interceptor trap should be washed through with hot detergent-hypochlorite solution. Make up a wash solution of detergent in hot water (10-12 litres at 60-70°C) With the addition of hypochlorite (250 ppm). In the case of branched lines, this quantity should be drawn in along each branch. Provision should be made to enable this to be carried out easily by providing at the end of each pipeline a cock or valve with a nipple suitable for attaching a rubber tube to suck the liquid from a bucket into the pipeline. When sucking the solution into the pipeline, allow air to be drawn in at the same time by withdrawing the tube from the solution at frequent intervals.

The air liquid mixture will achieve a scrubbing action in the pipeline and be more effective than a solid stream of liquid. Most interceptor vessels are about 18-20 litres capacity and must be emptied after each 12 litres of cleaning solution has been drawn into the plant. After washing the lines the drain cocks must be left open to allow the vacuum system to drain. Before re-installing the interceptor, check the rubber eals and observe if there are any small holes in the base of the interceptor where air leaks can occur. These can cause a carry over of liquid from the interceptor to the detriment of the vacuum pump.

- — carry out manufacturers maintenance instructions for pulsators Pneumatic self contained pulsators, if badly worn, can usually be made good by replacing worn components, ie. rubber diaphragms, valve slide and spring. Exchange reconditioned pulsators are available.
- — check for air leaks at MILK PUMP rubber non-return valve or shaft seals. With water in the receiver jar and under vacuum look for air bubbles in jar. If present dismantle and find cause.

Pulsator maintenance varies depending on type, (electronic, electro-pneumatic and pneumatic) and whether relay or self contained. Manufacturers instructions should be followed. Filters, air inlet ports and in the case of pneumatic pulsators, the valve and valve slides require cleaning.

A pulsator performs about 6 million operations a year and is therefore subject to considerable wear. A badly worn pulsator can usually be made to operate satisfactorily by replacing the worn components with new parts supplied by the manufacturer e.g. rubber diaphragms, valve slides and springs. It will normally be necessary after two or three years to replace pneumatic pulsators and pneumatic relays. Manufacturers may have an exchange service providing reconditioned pulsators.

Milk Pumps

The performance of a centrifugal milk pump can be seriously affected, even made inoperative, by air leaks at the rubber seal on the drive shaft or the non-return flap valve fitted in the pump outlet or delivery pipe. Where glass receivers are in use, test for leaks by putting approximately 4 litres of water in the receiver with the milk pump not working. With the receiver vessel under vacuum look for air bubbles rising to the surface from the receiver outlet. If a continuous stream of bubbles is seen dismantle the non-return valve and examine rubber flap for wear and distortion, renew if necessary. If leaks continue replace pump or dismantle and renew shaft seal, inspecting shaft for wear. It will be necessary to replace pump if there are signs of annular grooves on shaft.

Routine replacements

- Renew milking machine liners every six months, or 2,000 cow milkings.
- Replace long milk tubes every year.
- Replace long pulse tubes, rubber elbows and connectors every two years or when damaged.

Yearly

A full test of the milking machine must be carried out once

a year by an experienced technician as laid down in the International Standards; ISO 5707 "Milking Machine Installations - Construction and Performance" and ISO 6690 "Milking Machine Installations - Mechanical tests".

International Standards for Machine Milking Installation and Testing

In recent years the International Standards Organisation (ISO) a world federation of national standards institutes (ISO member bodies) has worked with other organisations for a standardisation of the terms and descriptions of the components used in machine milking and an acceptable performance of milking machine installations. It has drawn on recent research and devised a standard unifying the many national standards that existed. Furthermore a test procedure has been produced which can be used to assess the acceptability of milking machine systems and to check their performance at regular intervals after installation.

The International Standard, 5707, on Milking Machine Installations was prepared jointly with the International Dairy Federation (I.D.F.) and other interested bodies, ie., European Committee of Associations of Manufacturers of Agricultural Machinery (C.E.M.A.) and the International Committee on Recording the Productivity of Milk Animals (I.C.R.P.M.A.). It specifies the minimum performance requirements and certain dimensional requirements for the satisfactory functioning of milking machines. In addition it lists the requirements of materials, construction and installation. ISO Standard No. 6690 on Mechanical tests was produced at the same time with the object of providing test procedures for confirming that an installation meets the minimum requirements of ISO 5707. The following is a brief outline of items covered by the standards.

ISO Standard 5707

Vacuum pumps should be of adequate capacity to meet the operating requirements for both milking and cleaning and to provide a reserve. The reserve capacity is calculated for each milking installation. A formulae is given for calculating this, together with a formulae to calculate the effect of altitude on vacuum pump capacity. The installation of a separate vacuum system is recommended for ancillary vacuum-operated equipment. However, where this is not provided, an additional allowance for vacuum-operated equipment that does not operate during the static tests should be added to the effective reserve figure. Manufacturers are required to show the air consumption on all components. There are recommendations for the installations of vacuum pumps and exhaust systems. Facilities for measuring vacuum and air flows should be provided.

Vacuum regulators should be capable of controlling the vacuum pump used. They should be marked with the designed working vacuum level and with the air flow capacity at this level. A sensitivity standard has een introduced, together with standards for regulator leakage.

The acceptable drop in vacuum due to friction within the air pipeline (milking vacuum) is stated and recommended diameters of internal pipes are given for various air flow rates.

The interceptor capacity should be at least 15 litres. It should also be provided with an automatic cut-off and drainage facilitie. The inlet to and outlet from the interceptor should be the same diameter as the air pipeline.

A sanitary trap should be fitted to form a connexion between the vacuum system and the receiver vessel. Requirements for pulsation ratio are closely defined, specifying minimum 'collapse' and 'open' phases of the pulsation cycle.

Where milking pipeline systems are installed, a specific recommendation has been made for pipeline sizes to ensure that the vacuum drop does not exceed 3 kPa with all units working. The diameter is determined in relation to the total length of the line and the rate of milk and air flows. No risers are permitted.

Where recorder jars are used, the design should be suitable for in-place cleaning and recommendations are made in relation to the height of installation (ie. metres above sea level). To avoid any unnecessary vacuum drop in the milking system, attachments installed between a cluster and the milking pipeline (eg, recorder jar or bucket) should not cause a vacuum drop of more than 3 kPa at a milk flowrate of 3 kg per minute.

Detailed specifications and minimum internal dimensions are given for flexible tubes, ie, long milk tubes, 12.5 mm; short milk tubes, 8 mm; long pulse tubes, 7 mm except for bucket milking with alternate pulsation which is 6 mm; short pulse tubes, 5 mm and vacuum tubes 10 mm. There are also requirements for teat cup liners, shells and claws.

The standard also requires the installer to provide written instructions for operating and cleaning the equipment and also mechanical details of the system. The Mechanical Testing Procedures in ISO 6690 have been devised to enable static tests of the milking installation to confirm that it meets the requirements contained in ISO 5707.

Briefly this is done by first observing that the correct pipe sizes and method of installing and construction are used. Then with the mouthpiece of the liners sealed with bungs, the vacuum pump running and all components operating, confirm that the vacuum pump capacity is sufficient to give the effective reserve required to the size of installation. This is done by measuring the air flow and vacuum level with an air flow meter and vacuum gauge. Next a check is made that the regulated/controlled vacuum level is correct and stable. Then with a continuous vacuum recording instrument connected to the pulsation chamber of a teat cup assembly, record the pulsation characteristics and confirm that the results for each unit is satisfactory.

These results should be entered on a standard form and compared with the original test results. The installer of the installation should carry out the commissioning test with the records left on the farm for future comparisons.

References

McGee, Harold, "Milk and Dairy Products", *On Food and Cooking: The Science and Lore of the Kitchen* (in English), New York: Charles Scribner's Sons, 1984.

Merges, George, and Slade, Roger, "Dairy Development and Milk Cooperatives: The Effects of a Dairy Project in India", *World Bank Discussion Paper*, Washington, D.C.: World Bank, 1987.

7

Dairy Pasture Management

Livestock producers are resource managers, and as such are faced with the challenge of producing the, maximum amount of saleable product annually without destroying the long-term production potential of the farm or ranch unit.

Developing a Pasture Management Programme

A good sound pasture management programme is essential to meet the challenge. Good pasture management involves much more than simply turning stock onto forage. The ultimate purpose of pasture is to convert forage into a saleable animal product. To obtain maximum animal production, the needs of the animal as well as those of the plant (and their interaction of one upon the other) must be understood.

Many factors contribute to forage and livestock production. Good pasture management involves the control and application of these factors for maximum production.

Stocking Rates

The most important step in developing a pasture management programme is to determine the proper stocking rate. Overstocking pastures is a common mistake made by livestock producers. Overstocking reduces gains both on a per-animal and per-acre basis and destroys the long-term productivity of the pasture. Understocking a pasture produces maximum gains per animal, but production per acre is reduced. The ideal stocking rate is one that falls between the maximum production per animal and the maximum production per acre. The stocking rate of a pasture depends upon the:

— Vegetative cover of the pasture land
— Rainfall amount and distribution
— Fertility level and moisture-holding capacity of the soil
— Grazing system used (rotational versus continuous' grazing)
— Size and type of animal to be grazed

It becomes obvious when you consider the complexity of the above factors that the summer pasture programme must be tailored to an individual farm unit according to land base, livestock, climate and labor requirements. Assess your pastures at the end of each grazing season. The presence or absence of forage carry-over is the most important indicator of proper stocking rate and pasture condition. A 40 per cent carry-over of forage growth on native pastures is considered by researchers to indicate proper stocking rates.

The proper stocking rate for your farm or ranch can be determined by past experience, close observation of growth and a record of grazing performance on each pasture. Remember, the feed requirements of pasture animals increase as they grow, but forage production is not uniform and tends to decrease as the grazing season progresses. This important interrelationship between forage production and animal requirements must be considered in determining stocking rates.

Pasture Alternatives

The success of any livestock enterprise depends on an adequate supply of good pasture land. Pastures supply the cheapest source of gains for a livestock herd, and producers should strive to obtain maximum gains during the pasture season. Five pasture options available to producers are:

— Native pasture
— Tame pasture
— Annual pasture
— Complementary pasture

Consider the advantages and disadvantages of these five options in providing a source of feed for the total grazing season.

Native Pasture

Native stands are the major source of pasture for Manitoba cattlemen, making up 80 per cent of the total grazing area in the province.

Advantages:

— low input costs

Disadvantages

— low carrying capacities
— large acreage required
— low rates of gain
— short grazing season
— low production/acre
— limited management alternatives

Tame Pasture

Pasture improvement offers the greatest opportunity for increasing the income of livestock producers. With good management, land seeded to tame pastures can be the most profitable on the farm. Productive grass-legume pastures are essential for profitable production in today's livestock economy.

Advantages

— increased stock carrying capacities
— extended grazing season
— increased rates of gain

— increased beef production per acre
— smaller land base required

Disadvantages

— high input costs
— increased management requirements

Tame pasture, under a proper grazing system, can assure a continuous supply of quality pasture throughout the entire grazing season.

Annual Pasture

Annual crops are often overlooked as a valuable source of feed for the pasture season. Under normal growing conditions, a spring-seeded annual such as oats can be grazed four to five weeks after seeding.

Advantages

— increased stock-carrying capacity (compared with native pasture)
— excellent quality
— good palatability
— increased production per acre

Disadvantages

— annual cost input
— does not provide early spring grazing

Stocking rates on annual pastures should be heavy enough to prevent the crop from heading out. Vegetative growth will continue until a killing frost.

Complementary Pasture

A complementary pasture uses a combination of tame and native species, and the ratio used should provide for optimum forage growth and utilisation. The complementary pasture system is adaptable to most farms and ranches, and makes use of a small acreage of tame seeded pasture grazed in rotation with a large native acreage.

Advantages

- — increased stock-carrying capacity (over straight native pasture)
- — extended grazing season
- — increased gains
- — added flexibility in pasture programme
- — increased production per acre

Disadvantages

- — minor increase in inputs
- — increased management responsibilities

Grazing System

A grazing system can be defined as a schedule of when and where livestock are to graze during the pasture season. A successful grazing system is one that provides for the production of high quality pasture for the entire grazing season.

Continuous Grazing

Under this system, livestock are turned onto a pasture and left there for the entire grazing season.

Advantages

- — low labor requirement
- — minimum fencing required

Disadvantages

- — low stocking rates
- — poor utilisation of forage produced
- — livestock tend to overgraze selective areas, leading to weed invasion
- — system lacks flexibility
- — continuous grazing system is best suited to low-producing native pasture areas.

Rotational Grazing

This system involves dividing the pasture into separate pastures. The grazing livestock are moved from pasture to pasture throughout the grazing season according to forage growth.

Advantages

- — provides for higher stocking rates
- — more efficient use of forage production
- — provides flexibility
- — surplus forage production can be cut for hay
- — alfalfa can be maintained in the tame pasture stand

Disadvantages

- — additional fencing required
- — livestock must be moved according to schedule during the grazing season

A rotational grazing system is best adapted to seeded tame pastures or high producing native pasture. To obtain maximum benefits from rotational grazing there are three important rules to follow:

- — Move livestock to the next pasture before the animals have the opportunity to graze the grass twice during the same grazing period.
- — Graze the pasture while plants are in a leafy stage of growth.
- — Fertilise pastures regularly according to soil test recommendations.

Complementary Grazing

This grazing system makes use of a small seeded acreage of tame pasture grazed in rotation with a larger native pasture acreage. The complementary grazing system should be designed to use a combination of tame and native pasture in a ratio that will provide for optimum plant growth and utilisation of both the tame and. native forage species. The advantages and disadvantages of the complementary grazing system have been discussed under pasture alternatives.

Strip Grazing

This is an intensified system of rotational grazing used mainly for dairy cattle. The pasture area is grazed in strips. This is accomplished by setting a moveable electric fence across the pasture to allow animals access to only enough grass for one day. Each day the fence is moved forward to allow stock access to a fresh supply of grass.

Advantages

— provides for higher stocking rates
— assures daily supply of fresh forage
— increases production per acre

Disadvantages

— additional labor required
— requires uniform distribution of moisture throughout growing season

In both rotational and strip grazing, there is an advantage in separating producing and non-producing cows in the dairy herd. Producing cows should be the first to graze, followed by the non-producers as clean-up animals.

Mechanical or Zero Grazing

Under this system forage must be cut daily by machine and hauled to the livestock. As in any pasture programme, the forage should be used when it is in the early growth stage.

Advantages

— increased total forage production/acre
— no losses from tramping or fouling
— no opportunity for selective grazing
— less fencing required

Disadvantages

— high cost system

— high labor and machinery requirement

The additional cost of zero grazing may exceed the value of the increased production obtained. Data collected at the University of Manitoba indicates zero grazing is inferior to rotational grazing in terms of milk and butterfat production per acre.

Land base and herd requirements are equally important in deciding which system or combination of systems is best suited to an individual farm. Increased production can be realised by adopting a more intensive grazing management system. However, as management is intensified, the cost also increases. It is important to realise a dollar return from management. Farmers and ranchers must balance the additional cost of labor, fencing, fertiliser and machinery against profits when selecting a workable management system for their farms.

Livestock Management

Nutrition

The interrelationship of nutrition and disease control affect the profitability of livestock production; undernourished cattle are more susceptible to disease, and do not perform well either on a pasture programme or in the feedlot. As well, unhealthy animals usually require additional nutrition. Gradual changes from the winter ration to the summer diet of grass eliminate digestive upsets and their secondary side effects.

A mineral and salt feeding programme is as important on grass as in the feedlot. Amount consumed and type of mineral and salt needed varies throughout the province. A rule of thumb is free choice 1:1 stock mineral (e.g., in the Interlake high copper and zinc 1:1 stock mineral) with free choice cobalt iodized salt. For individual area recommendations, check with your local veterinarian, Agricultural Representative or feed dealer.

Implanting is not a replacement for good management, but it is a fast, easy procedure that can provide a slight edge in gains for yearling cattle on pasture. Several types of implants are available, the choice will depend on local availability, cost, and ease of application. Remember, for best results, implants require proper application plus adequate nutrition and good health.

Fresh, clean water is a necessity for efficient pasture gains. Water should be in plentiful supply, of easy access, and close to the grazing area. Dirty, stale and muddy water-holes not only affect weight gain, but are also conducive to health problems.

Vaccination and Immunisation

All pasture programmes should include disease prevention. Health needs vary from one region to another but most will include a seven or eight way Clostridial vaccine; I.B.R. (either intranasal or intramuscular); growth promotant; and fly control.

In some areas it may be necessary to vaccinate for B.V.D. and treat for parasites. These two procedures are optional and should be discussed with the local veterinarian in your area. If buying thin, unthrifty cattle for pasture, consider giving them injections of Vitamin A, D and E.

Sanitation and Hygiene

Basic sanitation provides a healthy environment and should eliminate many sources of secondary infection. This includes:

— Regular Housekeeping, i.e. proper storage of veterinary supplies; proper cleaning of needles and syringes after use; maintenance of handling facilities; and elimination of stress during handling.

— Good Water Supplies, ensuring that there is a clean, fresh supply, free of manure and other contaminating drainage.

— A Fly Control Programme using ear tags, oilers or dust bags will cut down stress on pasture cattle caused by biting flies.

— Quick and proper disposal of dead animals.
— Visitor Control; insist that all persons visiting the premises have clean boots, especially individuals who visit a large number of livestock operations (veterinarians, livestock buyers, salesman, government staff, etc.).
— Overgrazing means undernourished animals and the possibility of increased health problems.

Constant Surveillance

Unexpected problems can occur despite following all recommended management practices. Pastures should be checked daily if possible, as on-site observations will help to detect things such as foot rot, pinkeye, missing cattle, fence problems, water problems, etc. It also helps to have all cattle identified on an individual as well as herd basis. (tagging and/or Branding).

Proper preparation of cattle for pasture and care during the grazing season results in increased breeding efficiency, best use of pasture, minimum death rate and optimum returns.

Rotational Grazing

Ruminants such as cattle, sheep, and goats can convert plant fiber—indigestible to humans—into meat, milk, wool, and other valuable products. Pasture-based livestock systems appeal to farmers seeking lower feed and labor costs and to consumers who want alternatives to grain-fed meat and dairy products. The choice of a grazing system is key to an economically viable pasture-based operation.

Adding livestock broadens a farm's economic base, providing additional marketable products and offering alternative ways to market grains and forage produced on the farm. In addition, soil losses associated with highly erodible land used for row crops decline when such land is converted to pasture. Besides these benefits, rotating row crops into a year or two of pasture increases organic matter, improves soil structure, and interrupts the life cycles of plant and livestock pests. Livestock wastes also replace some purchased fertilisers.

Because ruminants co-evolved with grassland ecosystems, they can meet their nutritional needs on pasture. A profitable livestock operation can be built around animals harvesting their own feed. Such a system avoids harvesting feed mechanically, storing it, and transporting it to the animals. Instead, the livestock are moved to the forage during its peak production periods. Producers manage the pasture as an important crop in itself, and the animals provide a way to market it.

Reduced feed and equipment costs and improved animal health result from choosing species well-suited to existing pasture and environmental conditions. In most operations, a good fit between animals and available pasture provides more net income.

Some animals will produce acceptable meat with little or no grain finishing. Marketing these lean meats directly to consumers is an opportunity to increase profits. Skilled managers who can consistently offer high-quality forage to their animals, producing lean and tender meat, should consider pursuing this market.

Choosing a Grazing System

Continuous grazing, the most common grazing system in the United States, usually results over time in a plant community of less-desirable species. When livestock graze without restriction, they eat the most palatable forage first. If these plants are repeatedly grazed without allowing time for their roots to recover and leaves to regrow, they will die. Plants not eaten by livestock mature and go to seed. Thus, populations of undesirable plants increase, while preferred plants are eliminated, reducing the quality of the forage in a given pasture. Trampling and animals' avoidance of their own wastes further reduce the amount of usable forage.

Continuous grazing does, however, have the benefit of low capital investment, since few fencing and watering facilities are required. Because livestock are seldom moved from pasture to pasture, management decisions are simple. This type of grazing frequently results in higher *per-animal* gains than other grazing systems, as long as adequate forage is available to maintain high growth rates. But if pastures are overstocked, growth rates dwindle.

Rotational (or controlled) grazing, on the other hand, increases pounds of animal production per acre. How the system is managed influences the level of production, of course. In fact, management-intensive grazing (MIG) is another term for rotational grazing. This term emphasises the intensity of the management rather than the intensity of the grazing.

Management-intensive grazing (MIG) is grazing and then resting several pastures in sequence. The rest periods allow plants to recover before they are grazed again. Doubling the forage use on a given acreage is often possible with the change from continuous to controlled grazing. There is considerable profit potential for the producer willing to commit to an initial capital investment and increased management time. The producer can meet individual animal gain or gain-per-acre goals with sound management decisions.

Faced with low milk prices, the potential loss of price supports, and ever-rising costs, some dairy producers have changed to MIG to meet economic and quality-of-life goals. Some are providing cows fresh paddocks after each milking. Seasonal dairying—drying off the entire herd during times when pasture production is low— is often the next step, but it requires even more skillful management and may not be as profitable.

MIG can be used in many other operations as well. Cow-calf and stocker operations benefit from increased forage and higher-quality feed under MIG. Some graziers specialise in dairy beef or in raising replacement heifers for dairy operations. When MIG is used with sheep and goats, fencing must be excellent in order to keep the livestock in and the predators out. MIG offers the manager a wide range of options in terms of grazing intensity.

Changes in Grazing Management

When making a change in grazing management, a logical first step is an inventory of the farm's resources. An outline to help in this inventory process is enclosed. Another useful tool is an aerial map of the farm on which to mark fences, water supplies, and existing forage resources. Writing down farm and family goals in this process makes it easier to stay on course with management decisions. When a salesperson is applying pressure, for instance, it helps to be able to evaluate the cost of the product against some chosen goal.

Implementating rotational grazing requires subdividing the land into paddocks, providing access to water, adjusting stocking rates, and monitoring grazing duration. These decisions may seem overwhelming at first. Some of the enclosed materials offer information about setting up paddocks to fit the landscape, calculating stocking rates, and estimating forage yield and availability.

The change to controlled grazing will have impacts on the animals, the plant community, and the farmers. Livestock operators who have not monitored their livestock daily or weekly will feel the greater time demands. On the other hand, the need for harvested forages declines, resulting in less time spent making hay or silage. Purchased feed costs also shrink.

Economic benefits come from improved animal health and increased production. Research confirms lower feed costs and fewer vet bills on most operations making this transition. Actual figures vary widely, depending on the profitability and forage condition under the old system. As the new system is fine-tuned, feed quality improves, quantity increases, and management skills also grow. As a result, more animals can be raised on the same acreage, translating into more income for the farm.

It takes commitment to succeed in making the change to MIG, a system requiring more complex management skills.

Old ways of thinking will need to shift, as analytical and problem-solving skills develop. The new grazier's commitment will be tested by mistakes, unexpected weather patterns, and neighbors' attitudes.

Fencing and Water Systems

Rotational grazing requires additional fencing. High-tensile electric fencing is cheaper and easier to install than conventional fencing. Temporary as well as permanent electric fencing is available, and many producers use a combination of the two. This equipment offers flexibility in managing animal and plant resources.

Animals need to be trained to electric fences. Producers sometimes use a special paddock for introducing new stock into the system (fencing suppliers can furnish information). Once animals learn to respect the electrified wire, it becomes a psychological rather than a physical barrier.

Providing water is another capital requirement of rotational grazing systems. Experienced producers soon see the value of adequate water, and some regret that they did not invest more in the water system initially. Designing a water system for *future* expansion may be the best option for beginners with limited funds.

Many producers use pipes and portable waterers to create movable water systems and design permanent systems based on this experience. Flexibility in locating water within paddocks should be part of any final design, so the manager can control animal distribution and avoiding trampling around the water source.

Some paddocks have alleyways that give animals access to one water source from several side-by-side paddocks. However, the area around a permanent water source will suffer from heavy traffic. This heavy-use area tends to accumulate nutrients and is a potential source of parasites, disease, and erosion. (Many producers see the same problems in any location where animals congregate; e.g., shade trees and mineral sources.)

Heavy livestock traffic around ponds, springs, or streams can destroy vegetation. Piping water away from these sources or limiting animals' access results in higher-quality water for them, and it benefits wildlife habitat. Some producers report economic benefits from providing cool, high-quality water, though little research exists. Mineral blocks are typically placed near the water supply, but excessive use of the area can lead to the problems mentioned above. Placing the minerals away from water or other gathering areas helps redistribute the animals' impact and avoids overuse of any one area. Dispensing soluble minerals in the water is another alternative

Forage Growth

How much pasture area to offer animals and how long to keep them there are critical decisions for a successful grazier. These decisions influence the amount and quality of forage available throughout the grazing season.

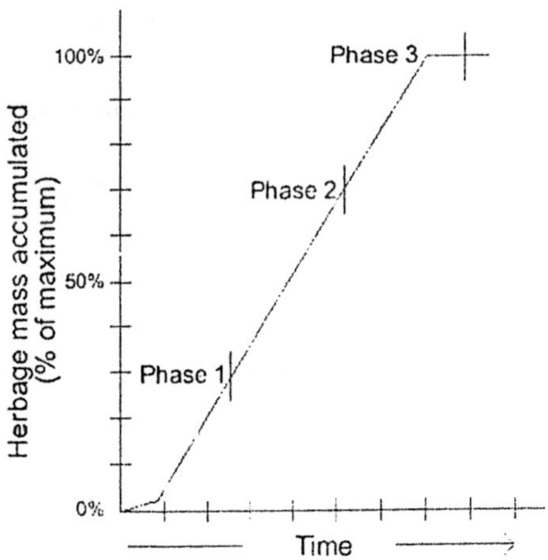

Figure 1. Forage Growth Curve

Figure 1 shows the natural progression of forage growth through three stages. Phase one is the first growth in the spring or the time required for regrowth after extreme defoliation. Photosynthesis is low because of the small leaf area available to capture solar energy.

During phase two, plants grow rapidly because leaf area is increasing. Toward the end of this growth phase, forage growth is near its peak, and it is of high quality. This lush and abundant forage is ideal for grazing.

The transition from phase two to phase three marks the beginning of reproduction and slower plant growth. Lower leaves begin to die as they are shaded out by those above. Plant resources are used for reproduction rather than more growth, and forage quality declines.

Managing Forage Growth

The grazier manages this forage growth-curve to keep pastures producing a maximum amount of high-quality forage. Decisions about moving animals from paddock to paddock are based on the amount of forage available, size of paddocks, and estimated seasonal growth rates. The number and nutritional needs of the livestock must also be figured into this balance. Additional information on these management decisions is included in the enclosures.

After each grazing period, if adequate leaf area is left for photosynthesis, plants quickly replace leaves lost without depleting root reserves. The animals are moved to fresh, succulent pasture before plants are overgrazed. Thus the plants and animals both benefit from good grazing management.

Many desirable plants, including legumes and native grasses, disappear from pastures that are not given adequate rest. Animals must be moved after three to five days, maximum, to prevent them from grazing these plants' regrowth.

If not removed from the area, livestock will preferentially graze certain forages and deplete root reserves, thus killing the most palatable forage species. Continuous grazing thus eliminates desirable species and maintains those that can tolerate repeated defoliation, such as tall fescue.

Management-intensive grazing encourages a wide variety of plants in the pasture. Plant diversity increases in adequately rested pastures. Plants adapted to the varied soil and moisture conditions of the landscape thrive in their microclimates. Animals can graze plants during their seasons of maximum palatability.

Livestock will, in fact, eat many weeds in their vegetative stage, some of which are good feed. By eating weeds such as dandelions, quackgrass, redroot pigweed, and lambsquarters when they are young and tender, grazing animals keep both annuals and perennials from going to seed. These plants have been shown to have feed values that compare favorably with oats.

Dairy or fast-growing meat animals will need energy or fiber supplementation at certain times of the season, depending on what they can graze for themselves. Since what livestock eat is different from a random profile of the plants in the pasture, forage samples or harvested forage tests will not exactly reflect true animal intake. It is, therefore, difficult for the manager to know whether protein or energy supplementation is economically justified.

Other than salt, the need for mineral supplements is likewise difficult to determine. If soil tests show micronutrients are missing, they can be added to the mineral mix. However, some may be present in the soil but unavailable to the plants. Adjusting pH often remedies this. While some consultants argue that missing micronutrients should be applied to the soil so they can be eaten as plant material, mineral suplements are often the most economical solution. Minerals not removed by grazing will cycle with other nutrients in the pasture as the years go by.

Seasonal Adjustments

Rotational grazing gives the livestock manager flexibility in responding to the changing forage supply. During periods of rapid plant growth, cattle are moved quickly through paddocks. Alternatively, if equipment is available or the work can be hired, excess forage can be harvested for feeding later. During periods of slow plant growth, delayed rotation allows plants in each paddock a longer time to recover after each grazing period.

Various strategies or specialised forages can delay having to feed harvested forages. In late fall, stockpiled fescue or other winter grasses can be strip grazed. Grain and stalks left in corn or milo fields after harvest, offered as strips, provide another source of good-quality feed into the winter months. Small grains, grown alone or with brassicas, are a third option in some parts of the country for extending the grazing season.

In some regions, providing excellent grazing through the hottest summer months is the biggest challenge. Native grasses, summer annuals, and interseeded legumes can offset this slump. However, the costs of establishment—in time and money—are justified only if the resulting increase in livestock production translates into sufficient profit.

Effects on the Animals

Multiple paddocks make access and handling easier. Cattle become easier to work when they see people as the source of fresh pasture. Managers who observe their animals frequently can identify and treat health problems in their early stages.

If just beginning an animal operation, the producer should choose a breed adapted to the climate and grazing system or pick individual animals with good performance records on pasture. Some types of animals, even within a breed, can better use high-quality forage, and others are better adapted to low-quality rangelands. Some tolerate legumes without bloating.

There is as much variation among individuals within the breeds as between breeds. To some extent, animals learn grazing skills. Therefore, animals that have been raised on pasture—especially those from a controlled grazing system—are desirable. In an established herd, culling animals that don't adapt is essential to achieving a profitable grass-based livestock system.

Management Intensive Grazing

Livestock producers can be viewed as "grass farmers." They produce the best quality and quantity of forage on a piece of land and harvest the forage with livestock. The level of productivity depends not only on hw muh foragis produced, but how efficiently it is harvested.

Grazing animals select the most desirable plants and avoid others. The composition of the diet they select will be more nutritious than the composition of the forage available. Selective grazing was essential to free-roaming animals prior to the fencing of the Great Plains. It allowed them to remain healthy and reproductive. Animals traveled in large herds over vast areas, which allowed forages to recover after grazing. This rest peiod ws vit to forage survival.

Today, however, livestock are confined to a given area of land for a specific length of time. Grazing management focuses on the relationship between animal numbers, amount of land, and time. The way producers use these three variables determines the efficiency of the forage harvest.

Contrary to conventional wisdom, it is not the result of too many animals. It is damage to a plant that reduces the plant's ability to functio normaly t occurs on a plant-by-plant basis. Damage is caused by the frequency rather than the severity of grazing. A plant can be overgrazed only when it is growing on stored energy.

Continuous grazing is the most widely practiced form of grazing. Its popularity is based on ease of use and possibly tradition. One pasture is used during the year or a particular growing season. The stocking rate is set to ensure the availability of forage during the growing sason. Ue stocking or over stocking will occur because of the variability in rainfall and forage production.

During periods of fast forage growth, consumption can not keep up with production causing poor forage utilisation. Some of the forage is allowed to mature and becomes less palatable. Animal performance will be good, however, because livestock can select the more palatable forages. During periods of slow forage growth, consumption exceeds production. Animals overgraze the more palatable orage anare forced to consume the less palatable forages. The result is poor animal performance and poor forage utilisation. The long-term effect of continuous grazing is replacement of preferred forages with less desirable forages or weeds.

Principles of Management Intensive Grazing

There are different scales at which we can discuss management-intensive grazing (MiG). One is the small scale of plant productivity and quality. At the other end of the spectrum is the large scale of the whole farm. In today's economy, agriculture faces challenges. The manager of a MiG forage-livestock system needs to study them carefully and identify solutions for the challenges facing the operation.

There are six major challenges facing managers of MiG systems and discuss some of the management strategies that can be used as solutions. The degree of implementation of a management strategy depends on the local farm, community, and market economics. These six challenges were chosen because of their impact on farm social, economic, and environmental sustainability. They are:

— Management direction

— Marketing

— Nutrient management

— Seasonal forage growth patterns

— Forage species

— Grazing management

Management direction

To direct the efforts of management and labor, managers need to identify what they want to accomplish through the farming operation. They may include personal aspirations of owners, managers, and workers; financial requirements and goals; and personal satisfaction and quality of life. Direction is based on a clear vision of what is to be achieved in the long term and how it will be achieved through short-term, intermediate, and long-term goals. The importance of visioning has been taught for over 3,000 years as noted in the quote: "Where there is no vision the people perish...". A more recent phrase for the need of vision is "begin with the end in mind". If a clearly defined vision and related goals are not articulated, then there is little likelihood that management or labor efforts will be directed in a cost-effective manner. Envision your destination and develop practical goals to draw the map to get there.

Marketing

If vision gives the destination and goals are the map, then marketing is the fuel for achieving the vision. If you put together a budget for an enterprise using average production, average costs, and average prices, most often the bottom line will be negative or break-even at best. To have a profitable MiG system, the manager has to be above average in some aspect of cost-effective production or marketing.

In West Virginia, pooled feeder calf marketing adds up to an additional $120 per calf sold. In the Midwest, this value may not be as high since feedlots are nearby and there is a stronger market, but good marketing practices will still be important. Livestock managers need to know what drives the market they are producing for and provide livestock with the attributes that add value to their product. In West Virginia, this is accomplished by pooling feeder calves in tractortrailer load lots, with calves having known genetic quality and being health managed under a certified quality assurance programme. In these pools, cooperating producers cull cows that do not produce calves meeting the pool requirements and use performance-tested bulls that are above average in performance EPDs.

A major part of marketing is to know your break-even price and the marginal value of your product. These two pieces of information allow you to determine if a management technology is likely to make additional money. An example of how to calculate the marginal value of feeder calf gain is given in Table 1, using average prices received by producers in the West Virginia Marketing Pools in 2000.

As Table 1 shows, a calf in the 500-pound weight class was worth $1.17/lb, but the value of a pound of gain going into the 600-pound weight class was not $1.17 but $0.36. When one buys a product to promote gain, the economics are different if the gain is worth $1.17 or $0.36 per pound. Table 1, calculate the marginal value of calf gain from reported feeder cattle prices by multiplying the midpoint weight of the reported weight breaks by their value. The difference in price per head for hundred 100 weight breaks gives the value of the marginal gain. For weight breaks of less than 100 pounds, adjust the value to a pound or hundredweight basis.

Table 1

Weight break	Midpoint weight	Average price/100 (lbs)	Average price/head lbs	Value of next 100 pounds added to calf
400-499	450	$124.00	$558.00	$85.50
500-599	550	$117.00	$643.50	$36.14
600-699	650	$104.56	$679.64	

Table 2 shows the value of different attributes of a feeder calf in the West Virginia market over a six-year period. A basic calf was worth $77.20 before adding weight, grade, or color. Adding 100 pounds to the calf added $57.52 so a 600-pound calf added $345.12.

Table 2. Factors that affected the value of a steer feeder calf in the West Virginia market over a six-year period (1990-1995)

Component	Model	Average	Range Covering 66% of Observations
Calf ($/hd.)	Basic calf	77.20	± 27.16
Weight ($/hd. /cwt.)	Sale weight	57.52	± 5.93
Grade ($/hd.)	M1	49.29	± 7.17
	L1	44.51	± 6.85

	S1	5.58	± 7.35
	LM2	0.00	± 0.00
Breed ($/hd.)	Black white face	25.72	± 8.67
	Angus	19.27	± 5.94
	Charolais cross	17.35	± 6.36
	Continental cross	11.60	± 7.60
	Hereford cross	0.00	± 0.00
	Hereford	-5.80	± 6.52
Date ($/hd./ day)	Late (after Oct 1)	-0.52	± 0.26
Lot size	Early (before Oct.1)	0.52	± 0.26
($/hd./hd.)	No. head in sale lot	0.81	± 0.05

If the calf was a M1 Angus calf, you added another $49.29 for the grade and $19.27 for the breed, bringing the value of the calf up to $413.68 or $82.74/cwt. Then there were additional premiums for the date of sale or how many cattle were in the lot of cattle sold.

Understand your market and use enterprise budgets to determine and control the marginal costs and profitability of management strategies in your MiG system.

Nutrient management

Good management of plant nutrients is needed if a MiG system is to be economically or environmentally sustainable. Soil fertility and fertiliser (including manure) management can make the difference between a soil producing only 30% of the soil's yield potential or 100% of its potential. The fertiliser value of manure and urine passed to the soil by a 1,000-pound cow is worth $100 to $130 per year. If the animal's manure and urine are not spread relatively uniformly where they can grow grass, their value is lost. The replacement value of these plant nutrients represents 20-25% of the value of a weaned calf. Nutrient flows in a MiG system can be managed by proper placement of fencing and a watering system and by shade management when shade is necessary.

Legume management is another tool that provides nitrogen at relatively low cost and risk compared to purchased nitrogen. Nitrogen fixation by clovers or alfalfa replaces the need for commercial nitrogen in cool-season pastures. Legumes also increase forage intake compared to nitrogen fertilised grass, increasing yearling cattle growth by 25% or more and increasing dairy cow milk production by 6-10 pounds per day. Table 3 shows data from Virginia where clover used in place of 200 pounds of nitrogen per acre per year gave improved steer growth and a small reduction in grazing days per acre due to the increased forage intake, but about the same gain per acre.

Table 3. Legumes used with a cool-season grass increase forage intake and animal gain per head at a lower stocking rate with little loss in gain per acre compared to the grass alone fertilised with 200 pounds of nitrogen/per acre per year

Pasture	Days/acre	ADG	Gain/acre
OG-Clover	257	1.28	329
OG-200#N	311	1.07	333
TF-clover	303	1.02	303
TF-200#N	403	0.91	367
BLG-clover	258	1.21	312

A good MiG tool is to base a cool-season pasture on legume-fixed nitrogen and use strategic nitrogen applications when they are economically justifiable.

Seasonal forage growth pattern

Pasture systems based on cool-season grasses and legumes have a flush of growth in the spring when soil and air temperature and soil moisture are optimum for their growth. In the summer as temperatures increase and soil moisture decreases, these forages have a growth rate onehalf to one-third of the spring growth rate.

A cool-season pasture system needs a "buffer" to balance forage production and quality to the animal nutritional requirement. If the manager does not develop a buffer, Mother Nature will provide one of her own. If there is no buffer and a pasture is stocked for average midsummer forage growth, the animals will waste about 50% of the spring pasture, representing about 25% of the total forage growth for the year. The animals will also be short of feed at the end of summer, which can reduce milk production or growth.

If late-spring growth was adequate, then the area might be hayed. We can broaden the definition of buffer to include other management strategies that enable the manager to match forage availability to animals' needs. Strategies that act as buffers in a MiG system include:

— Harvesting first-cut hay on some paddocks and adding these paddocks back into the grazing sequence in midsummer to provide more acres of pasture.

— Varying the stocking rate by shipping some livestock in early summer when pasture growth slows down.

— Using warm-season grasses that do not need to be grazed until late spring or early summer in selected paddocks.

— Wasting some forage in the spring.

— Allowing animals to lose body condition in late summer.

— Feeding supplemental forage, forage substitute, or grain when forage growth slows down.

— Strategically applying nitrogen fertiliser to increase fall growth.

Depending on local economics, some of these strategies are better alternatives than others. Using a buffer in a MiG system can increase the effective productivity of a soil by 25%. Use a grazing system buffer to increase the efficiency of forage utilisation and use stockpiling or other fall forage production strategies as fall buffers to extend the grazing season by one to three months or more.

Forage species

In most cases, the best economic approach in forage production is to manage well what is in the pasture. It is often amazing the response an old, worn-out pasture gives once it is taken care of by rotational grazing, a buffer, improved soil fertility, and legume management.

Soil types can range in yield potential from 2 to 5 tons per acre or more. These two extremes can be found in soils located next to each other on a farm. Fencing should be laid out so that soils having similar drainage and yield potential are within a paddock. This will assist in grazing management and the development of more uniform species and forage growth within the paddocks.

If improved management does not increase pasture productivity to a level appropriate to the soil's potential, it may be worth a new seeding. The species mixed for a new seeding should be ones adapted to the site's climate, soil drainage, and grazing management. For soils having potential hay yields of 2.5 tons or less under good management, there is little likelihood of economic improvement beyond a good grass-clover system.

If the grass is endophyte-infected tall fescue and clover and grazing management is not proving adequate to meet the farm's production and economic goals, then eradication of the infected fescue and establishment of an endophytefree or endophyte-enhanced tall fescue may be advisable. When used with ladino and red clover, these new fescues have the potential to be the foundation of good MiG systems for meat and milk production.

Grazing management alone may not bring in the plant species most adapted to the extremes in soil drainage. On deep, well-drained soils, seeding alfalfa may be a good practice. An alfalfa-grass mix makes good first-cut hay and provides a good buffer for summer grazing. Since alfalfa can exploit moisture in deep, well-drained soils, total yield may be increased. The hay buffer adds to grazing flexibility and system efficiency. The grazing management of alfalfa paddocks should be different than grassclover paddocks to meet the needs of the alfalfa. Wet sites may not have a natural seed bank of alkaloid-free reeds canarygrass so seeding this species may greatly increase forage production and improve paddock trafficability in wet weather.

Two cool-season grass species grown on a site, under management that both are adapted to, will seldom differ in yield by more than 20%. Varieties within a species seldom differ in yield by more than 10%. However, adding a nitrogen-fixing legume such as clover or alfalfa to a grass mix, where no commercial nitrogen is used, can increase total forage production by 100%.

If the plant species in a new seeding are not adapted to the soil and grazing management used, they will not survive and yield of the planted forage species may be zero. Plants arising from the soil seed bank will then invade the site. In the best case, the site may revert to the original plant community with only a loss of money. If there are major weeds that were controlled by the original plant community and grazing management, the site may be worse off and money may be lost.

The plant species found in a pasture are the ones adapted to the soil drainage, fertility, and current grazing management. Changing the grazing management and soil fertility may change the plant community to a more productive, higher quality one. When reseeding, use a mixture of species adapted to the site's soil drainage and grazing management.

Grazing management

Fences and water are the tools for controlling grazing management. Movement of livestock in relation to pasture growth or defoliation controls the timing and intensity of grazing, which is functional grazing management. Stocking rate is used to balance annual forage requirement to annual forage production and utilisation. Buffers are used to balance monthly forage requirement to monthly forage production and utilisation.

The foundation of proper rotational grazing is timing and intensity of plant defoliation to meet the nutritional needs of the animal and maintain plant health. At times, preference may be given to the needs of the animal. At other times, we may make the animal work for us to accomplish a plant management goal. The effects of grazing, rotation rest interval, and grazing intensity (grazing pressure) in MiG can mean the difference between 50 and 100 percent production relative to the soil's potential.

The term "intense management" is sometimes used to mean the number of paddocks in a grazing system. "Intense grazing" should be used to refer to how close or short the animals are allowed or forced to graze a paddock before they are moved to a new paddock. Some species respond well to close grazing while other species do not (native warmseason grasses). It is possible to mismanage 20 rotationally stocked paddocks as much as one continuously stocked pasture.

The management-intense part of MiG is based on:

- — Knowing the defoliation timing and intensity preferred for a given plant community in a pasture.
- — Controlling the defoliation timing and intensity of the pasture by moving animals between paddocks in a rotationally stocked system or by controlling the height of pasture by changing the stocking rate in a continuously stocked pasture.

References

Adams, Don et al., *Matching the cow with forage resources*, Rangelands. April, 1996.

8

Economics of Dairy Nutrient Management

There are several different types and combinations of nutrient management systems. Most systems accommodate six basic functions: (1) production, (2) collection, (3) storage, (4) treatment, (5) transfer, and (6) utilisation of waste. For a specific system, some of these functions may be combined, repeated, eliminated, or rearranged. Nutrients produced in milking parlours and confinement areas must be collected. Storage is the temporary procedure to contain the nutrients. The storage facility is the tool that gives the manager control over the scheduling and timing of the system functions. Treatment is a process designed to reduce the pollution potential of the nutrients, including physical, biological, and chemical treatment. Transfer refers to the movement and transportation of the nutrients throughout the system. Utilisation includes recycling reusable waste products and reintroducing non-reusable waste products into the environment. Agricultural wastes may be used as a source of energy, bedding, animal feed, mulch, organic matter, or plant nutrients. Properly treated, they can be marketable.

Alternative systems are created when different methods are used for any of the waste handling system components or when any of the components are rearranged or modified. Generally, dairy nutrient management can be classified into three systems—solid, slurry, and lagoon.

Solid nutrient management systems are commonly used in smaller operations (less than 100 cows) with bedded loafing barns or stanchion stalls. These systems minimise the volume of manure that is handled. However, a separate facility is required for liquid milking center waste. Manure with 75 to 80 percent moisture content can usually be handled as a solid. Manure at this moisture content has a consistency of peanut butter. Twelve pounds of bedding per 100 pounds of fresh manure (about 4 pounds of dry straw per cow per day) is needed to permit dairy manure to be handled as a solid. These systems require scraping devices, loaders, manure storage, and manure spreaders.

Slurry nutrient management systems maximise recovery of plant nutrients from waste and are often used where geologic conditions are unsuitable for a lagoon system. Compared with solid nutrient management systems, slurry systems increase the volume of manure handled because water content is higher than in solid waste, but they allow the manure to be handled as a fluid. Manure with 90 to 96 percent moisture content can usually be handled as a fluid, but may require special pumps. These systems require slurry storage, earth basins, scraping devices, pumps, and perhaps tank wagons.

Slurry systems require more land for application than do lagoon systems because more nitrogen is retained. Lagoon systems are favoured by many dairies because they have lower cost relative to other systems. Solids separators are used to reduce solids buildup in the lagoons and drastically reduce the frequency of lagoon dredging. Lagoons are generally preferred where flushing is desired and where a significant amount of lot runoff must be contained. Lagoon systems handle highly diluted waste (96 percent or more water) that can be pumped through irrigation flushing equipment if excessive straw or fibrous material is not present.

For conventional pumping, two gallons of water must be added to dilute a gallon of fresh manure to 96 percent. Of course, this greatly increases the volume of material to be stored and transported. Most lagoon effluent is more than 99 percent water. These systems require pumps and irrigation equipment. The latter are generally sprinkler systems and may be either stationary, handcarried, or moving systems.

Most operations with fewer than 100 dairy cows use some form of solid nutrient storage. Use of methods for storing manure in a liquid form increases with herd size. With the slurry method, manure is stored as a thick liquid in a pit under the barn floor or in a tank or earthbasin until it is applied onto land. These are most often used as short-term transfer tanks. With lagoons, either anaerobic or aerobic, manure is diluted with water, often from flush systems and milking parlour wash water.

Slurry systems are more common than lagoon systems for herds of fewer than 200 cows. Both systems are equally popular among producers with 200 or more cows. Over 90 percent of herds with 200 or more cows have some type of liquid manure storage. Since evaporation reduces total lagoon volume more than slurry volume, especially in more arid parts of the country, it is not surprising that lagoons are most common in the western United States. Producers with liquid manure systems in the Midwest and Northeast often prefer slurry systems over lagoons.

Principles for Selecting Nutrient System

Since nutrient handling does not generally produce significant revenues, impact on profitability can be measured in terms of net cost. Net cost is the cost of owning and operating the system minus revenues from the sale and/or value of distributed nutrients. The system with the lowest net cost is the most profitable.

In addition to economic considerations, factors such as environmental impacts, animal characteristics, facility investments, and nutrient distribution area must also be considered in selecting an appropriate nutrient-handling alternative. For example, the most economic alternative may not be relevant if soil conditions or limited land area make it infeasible.

Manure Production and Storage Requirement

For each nutrient handling system, the first step in estimating the cost of dairy nutrient management is to determine the nutrient storage requirements. Storage requirements depend on several factors, including the quantity of manure produced, bedding, wash water, slab runoff, and rainfall. Wash water comes from preparing cows for milking as well as from cleaning milking equipment, parlour, and holding area. Slab runoff refers to the precipitation collected and stored from confinement areas and slabs. When a separator is used with the lagoon system, separated solids do not enter the lagoon. Solids are generally not separated for the liquid tank system or for two-cell holding ponds.

For the lagoon system, total storage requirements must be sufficient for the total volume of waste, solids accumulation, normal precipitation less evaporation on the lagoon surface during the storage period, precipitation on the lagoon surface from a 25-year, 24-hour storm, and a margin for safety (freeboard). Solids separated prior to entering the lagoon must be stored until distributed, composted, and/or sold. Frequently, lagoons are designed to include outside runoff from watersheds. For such, the runoff volume of the 25-year, 24-hour storm must also be included in the storage volume.

Rainfall and evaporation in the major dairying areas of northwest Washington vary greatly by location and season. For a half-year storage period, the size of the nutrient storage facility is computed to accommodate winter conditions. During this period about 65 percent of annual rainfall is received. Annual rainfall is presumed to be 40 inches and evaporation is considered to be one inch per month during the winter storage period. A 24-hour, 25-year rainfall of 4 inches is also accounted for.

Investment Costs

Lagoon System

The storage lagoon is the most basic component of the lagoon nutrient management system. It is a treatment facility for slurry and liquid waste. While it can also be used to temporarily store all forms of nutrients, subsequent removal of solid and semisolid nutrients can be difficult and expensive. Location is important. The lagoon should be located as far as possible from houses and public roads and downwind so that prevailing winds carry odours away.

Lagoon odours can be objectionable at distances of ½ mile and detectable at distances of a mile or more. The lagoon should be located as close to the nutrient source as possible. If the lagoon is downhill from the nutrient source, gravity can transport the waste. Where possible, the lagoon should be located over impervious soil so that the bottom and sidewalls don't require sealing. Lagoons on many soils require sealing with liners, clay, or soil cement. Sealing may also be accomplished biologically. Animal nutrient solids are a good sealant in many soils, but this process takes time. Clay or soil cement delays leaking while biological sealing is developed. Membrane sealing (plastic, vinyl, rubber, etc.) is positive and effective, but it is expensive and difficult to install.

Common methods for transferring liquid dairy manure to storage include gravity flow, large piston pump, pneumatic pump, and centrifugal chopper pump. A piston-type pump provides convenient transport of manure to a storage structure. A key factor in the design of any liquid storage structure is provision for agitating the material prior to irrigating or loading the tank spreader. Without complete agitation, solids will accumulate in the structure and reduce storage capacity.

Irrigation equipment has been adapted for application of liquid manure and wash water on cropland. The primary concerns are to apply the nutrients at agronomic rates on cropland that has need of the manure nutrients and to apply them in an environmentally acceptable manner. The use of manure or wash water for "true" irrigation is seldom accomplished because of the relatively small volume applied. Those who desire to irrigate in addition to spreading manure must be certain of an adequate supply of water.

Pipelines used in nutrient management systems can be of the same type and general design of those used in normal irrigation systems. Because of the corrosiveness of the wash water, however, underground pipelines should be constructed of plastic or other non-corrosive material. Flushing pipelines and other nutrient-application equipment with clean water is recommended after each use and definitely before storage.

Dairy farmers face two options when they apply manure to cropland. One is to hire a custom irrigation system, which typically costs about $100-$125/hour. The other (which provides more flexibility in controlling amount, form, timing, and placement of dairy manure nutrients) is to purchase an irrigation system. A travelling gun irrigation system is most often purchased and is the option included in our cost calculations. Injection systems are increasingly being used near urban and environmentally sensitive areas in an effort to reduce odor and nutrient volatilisation and to utilise manure nitrogen more efficiently, near the roots of the plants.

Additional equipment is needed if the farmer doesn't hire a custom irrigation system. Equipment typically required includes a big gun sprinkler, irrigation pump, and pipe. Such equipment can add 25 percent or more to the nutrient management investment costs. Although expensive, purchasing the irrigation system reduces annual operating costs (to be discussed later) sufficiently to be economical for each of the herd sizes considered in this study. It may also be necessary in areas where custom irrigation systems are not available for hire and where equipment cannot be shared.

Additional equipment is required to load and possibly distribute dairy manure and bedding solids. Many dairies that separate solids develop markets for raw or composted material that fully cover the cost of managing the solids. Consequently, no additional equipment or operating costs are calculated here for managing solids. Costs and returns from composting the solids are developed in a later section.

Liquid Manure Tank System

In locations when a lagoon is not feasible because of geological or other conditions, a liquid tank system is often selected. The liquid tank system considered here is a cast-inplace, in-ground, concrete-covered storage tank with 180-day capacity that is loaded by gravity. Above-ground tanks would need a mechanised pump for loading manure into the tank. The primary investment cost of the liquid tank system is for construction of the manure tank. Typical construction costs are about $120,000 per million gallons of storage capacity.

Equipment requirements include a manure scraper, tractor, and a 3,000-gallon tank wagon pulled by a 100-horsepower tractor. This permits nutrients to be distributed within a 10-day period. Other equipment needed includes agitating and loading pumps (used to agitate the slurry in the tank while pumping from storage tank wagons), an open-impeller, and an irrigation reel or manure injector system.

Annual Fixed Costs

Annualised fixed costs range from 9.63 percent of the investment in the storage slab to 16.25 percent for equipment. Even for the largest herd size, fixed costs per cow are greater with the liquid tank system than for most herds with a lagoon system.

Annual Operating Costs

Operating costs for solids separation include power and labour which vary by herd size. A wage rate of $14/hour and a power rate of $20/hour are figured to compute the annual application cost when the irrigation system is owned. Per cow, these costs range from $47 per cow for 250-cow herds to $32 per cow for 3,000-cow herds. When the irrigation system is custom hired at $100 per hour, they range from $97 per cow for 250-cow herds to $70 per cow for 3,000-cow herds. Annual operating costs for the liquid tank system are a little lower than those for the lagoon system with a hired custom irrigation system. The annual operating cost per cow ranges from $87 for 250-cow herds to $69 for 3,000-cow herds.

Dairy Nutrients for Plant Production

Field Application of Manure

Dairy manure nutrients can help to build and maintain soil fertility. Also, it can improve tilth, increase water-holding capacity, lessen wind and water erosion, improve aeration, and promote beneficial organisms. In addition, when wastes include runoff or dilution water, they can supply moisture as well as nutrients to crops.

Proper manure application to fields is not only an indispensable part of the nutrient management, but also a critical step to prevent surface and groundwater contamination. Once manure is applied, it must remain on the field until it is absorbed by the soil. If manure moves beyond the targeted field, it becomes a pollutant. The extent to which manure is kept on targeted fields depends on the application method. For instance, big gun applicators provide the least control and accuracy for liquid manure application while tank-type spreaders and injector systems give the most control. Proper management, planning and vigilance during application keep manure in the desired target area and out of streams and ditches.

The soil infiltration rate measures the soil's capacity to absorb the liquid when manure is applied to a field. This rate depends on soil type, amount of solid material contained in the manure, speed and duration of application, and soil compaction. Existing soil moisture at the time of application affects the total amount of liquid manure that can be applied. If manure application exceeds the soil's infiltration rate, a portion may run off and pollute adjacent surface waters. Manure solids can also seal the soil surface causing infiltration to slow or stop.

The timing of application should also be considered when making the application plan. The best time to apply manure for crop fertiliser is spring and early summer when growing crops need the nutrients. At soil temperatures above 40°F, some of the applied manure nitrogen converts to leachable nitrate in the soil. When manure is applied to warm (60°F or higher) moist soil, it converts to nitrate in several weeks. Soil temperatures in the fall are still high enough to provide ideal conditions for converting manure nitrogen to nitrate.

Computation of Dairy Nutrient Value

On most dairies, operators use nutrients to reduce costs, or even achieve an economic return, since dairy manure can generally be used as a fertiliser and soil conditioner. Dairy nutrient applied to land has a value measured by the fertiliser nutrients of replaced commercial fertiliser and/or for the increased production of plant growth. Estimating the value of dairy nutrient requires several steps. Nutrient content of the manure must be estimated. Total nutrients are then adjusted for separated solids removed and for losses during storage and application. The value of the manure for fertiliser is computed based on the prices of fertiliser nutrients. The minimum acreage on which dairy nutrients can be distributed without environmental risk is also computed based on the crop, yield, and limiting nutrient.

The starting point for all calculations is to estimate the total nutrient content of the manure. Nitrogen, phosphorus and potassium are the major nutrients in dairy manure that are considered in computing an economic value. It is important to determine the predominant nutrient(s) that control planning and implementation of dairy nutrient application to promote crop production and environmental protection.

The total amount of these nutrients depends on nutrient production in the excreted manure, wash water production, nutrients in wash water, and storage period. Next, nutrient reductions from separated solids not distributed to the land and from storage and application losses are deducted from the total nutrient production. Nutrient losses from dairy waste can be grouped into three general categories—those that occur during storage, during application to the soil, and after incorporation. Nutrient losses from manure during storage and application vary widely and depend on climate and management, including methods used for collection, storage, treatment and application.

Timing of nutrient application is critical to conserving the nitrogen in the manure. Volatilisation losses increase with time, temperature, and wind, and decrease with humidity. To minimise volatilisation losses, manure should be incorporated before it dries. There are additional N losses through volatilisation (5-15 percent N loss) and denitrification (10-30 percent N loss) that are dependent on the time and method of application and on the soil drainage class.

Net Cost Computation

For the lagoon system, no costs are included for managing the separated solid material and no value is included for its application. The net annual cost of the lagoon system for farms hiring a custom irrigation system ranges from $97/cow for 250-cow herds to $25/cow for 3,000-cow herds. If dairy farms purchase their own irrigation system rather than hiring equipment to apply manure, the net annual cost is lower for all herd sizes ($79/cow to $2/cow, respectively).

The net annual cost for the liquid manure tank system is higher than for either lagoon system at all herd sizes. Net annual cost ranges from $167/cow for the smallest herd to $65/cow for the largest. These figures further document why the lagoon system is preferred unless soil or geological conditions preclude it. This conclusion is not sensitive over a wide range of interest rates. Additionally, since fewer nutrients are lost with the liquid tank system, additional acreage and/or higher crop yields are required to appropriately utilize nutrients.

Composting System

A composting system is a modification of a conventional or solid manure handling system in which a composting treatment process is applied to the manure. Composting is becoming more popular as a method of handling various agricultural wastes. Three factors may induce dairy farmers to consider composting: severe environmental constraints on traditional nutrient management procedures, increasing cost of handling dairy manure, and economic potential of composting. Although these factors encourage consideration of composting, there are several potential tradeoffs farmers must also consider. They include additional equipment and labour costs, land and improvements required to produce compost, composting technique, management required to assure high-quality compost, scale and scope of operation, and market for the final product.

Thus, the objectives of this evaluation of composting include:

1) Identifying alternative methods of composting.
2) Describing economic opportunities for composting solid dairy waste.
3) Determining investment, operating costs, and likely returns from composting.
4) Identifying the market potential for dairy compost.

Composting Methods

Various methods are available to produce compost. They include passive windrow, turned windrow, in-vessel/channel, extended aerated static pile, and vermi-composting. These methods vary greatly in the quality and consistency of compost produced, investment required, and operating costs.

The passive windrow method turns windrows with a loader, is relatively simple and cheap, but produces the lowest and least consistent quality of compost. When implemented on the dairy, this approach may not require any additional equipment or investment unless drying pads, runoff prevention measures, or covered space are needed. The volume of materials this method can handle range from a few hundred to several thousand cubic yards per year. This method of composting minimises new investment and requires a relatively low level of management intensity.

The turned windrow method requires more capital and labour than the passive windrow method, but the expenditures may still be reasonable because most dairy farmers already own some of the necessary equipment. Farmers may manage windrows at a moderate level of intensity by purchasing a specialised windrow turner. This method requires a moderate amount of labour and may require an investment as small as $10,000. For larger operations, a fully integrated, self-contained windrow turner costing about $200,000 greatly reduces labour costs. Final product quality is high and the composting period is short. For these reasons, the turned windrow method is currently the most popular method for on-farm composting.

The in-vessel/channel composting method requires little labour, product quality is high, and space requirements are small. The extended aerated static pile method requires a system of perforated PVC pipe covered with a layer of shavings and topped with about 8 feet of fresh manure solids and then covered with pre-composted solids. Air is forced through the pipes at variable rates essential to maintain consistent composting temperatures. However, these methods are not popular for most on-farm operations since they require investment in very expensive equipment and skilled labour.

Vermi-composting produces the highest quality compost and in some cases, can be the least expensive method. It requires little equipment and labour. The major requirements are a large amount of covered space, the means to move the materials (a turner or front end loader), and screening equipment. This method tends to be used only on a small scale.

Composting Merits

Composting converts nutrients to a more stable form, adds humic acid to the soil, increases beneficial soil organisms, improves soil tilth and aeration, reduces raw manure odours, and reduces reliance on synthetic fertilisers. Although compost is not usually marketed as fertiliser, it can add nutrients to the soil. Compost users include home gardeners and landscapers as well as farmers and local governments.

Most compost from agricultural waste is currently being used directly by the farm or local government composter (e.g., for easement plantings) or is sold in bulk in many locations for prices near $10 per cubic yard. In Washington, final compost is often marketed at $12 or more per cubic yard F.O.B. The price of compost depends on the amount purchased, quality, promotion, packaging, and associated services. In addition to the potential revenue from compost, it is frequently preferred for environmental reasons.

Manure used as compost quickly breaks down, provides slow release of nutrients, has less odor, may require less acreage for application (depending on soil nutrient load), and has excellent benefits for soil. In addition, waste disposal fees of $50 to $100 per ton have become common, so revenue can sometimes be generated by charging disposal fees.

Costs of Composting

Depending on the scale of operation and the technology adopted, initial outlays for planning, permits, site preparation, and investment in equipment and the site can vary greatly. Initial outlay can range from a few hundred dollars to hundreds of thousands of dollars. To determine net benefits or costs of composting, several factors must be considered - quantity of waste, land available for the compost facility, market for compost, and transportation costs.

Costs depend on the quantity of manure composted. Many farmers compost several thousand cubic yards of material without significant additional costs. However, when larger volumes of waste are composted, land, labour and capital investment can be substantial.

Land that can be devoted to composting will influence a farmer's decision on whether to compost. At least one acre of unused or underutilised land with suitable slope, drainage, and access is required for the composting facility. Concrete slab and cover may be important for efficient composting. Compost leachate must also be contained or filtered to avoid water contamination. Depending on the technology used, one acre can accommodate 2,000-10,000 cubic yards of compostable material per year. Larger investments in equipment or technology can substitute for scarce land.

The market outlet for finished compost is critical. Compost of consistent high quality can generally be sold easily and profitably but requires careful management to assure desired carbon to nitrogen ratio, temperature control, and bacterial content. More capital investment in equipment, such as screens and monitoring equipment, may be needed to improve the quality and consistency of the final product. Marketing costs are also frequently required when farmers sell compost rather than applying it on their own farms. However, if they plan to add all or most of the compost to their own soils, they can simplify their compost systems and avoid much of the expensive extra processing since it adds little value to the compost for farm application.

Transportation costs can be substantial and warrant specific consideration. They include the cost of transporting manure to the compost site and then transporting final products to market or to the land where the compost will be applied. Transportation costs may increase substantially if solid manure and bedding from several dairies is transported to a central composting facility to take advantage of economies of size. Carefully balancing transportation costs and economies of size can help minimise costs per ton of composting and make possible the efficient utilisation of expensive fixed investments such as specialised composting equipment or land.

Costs of a specific compost system also depend on additional variables, which vary from farm to farm. Such variables include labour cost, fuel price, land value, equipment investment and maintenance cost. Because various combinations of land, labour and equipment can produce desirable compost using different technologies and management systems, the farmer has several options for using existing resources in a cost-effective way.

A loader is required for the passive method and a windrow turner and screen are needed for the turned windrow method. A screen is used to separate materials of different sizes and shapes and improves the quality of the compost for sale or use, but it is not necessary if farmers choose to apply compost to their land rather than selling it. In farm composting systems, the screening is nearly always performed following composting.

Although adequate tractor and loader capacity and storage space may be available on smaller dairies to manage a passive-windrow composting system, their full cost is considered here. The additional storage slab for on-farm composting is about twice the area required to store solid waste without composting. For the centralised composting facility, the full investment in the storage area is included (i.e., three times what would have been required for a single dairy of the same size to store solid waste without composting). Because of the large number of hours required with a passive-windrow system, one tractor and loader is required for every 1,000 cows.

Benefits of composting

Composting can be introduced for many reasons. It can develop a marketable product from waste, improve manure application management, provide soil conditioning and/or a bedding substitute, and reduce the risk of pollution and nuisance complaints. Since concerns about traditional manure utilisation methods are increasing, both dairy farmers and society in general are searching for alternative manure utilisation methods. This could create an opportunity for farmers to collect processing or tipping fees by composting offfarm waste materials, such as municipal ward waste, horse stable bedding, or vegetable processing by-products.

In order for some manures to compost properly, they must have sufficient carbonous materials included. Off-farm waste materials are often ideal for that purpose. Regulations vary, but after meeting those that require on-farm use, sales of the composted material can be lucrative. Economic benefits from composting may also occur due to reduced annual operating costs for manure application and increased revenue from selling compost.

Although there are many benefits to on-farm composting, composting requires equipment, labour and management.

Composting may also require additional storage space for raw materials and the final compost product. Weather is an important factor to consider before starting an on-farm composting program. Cold weather and heavy precipitation greatly affect the composting process and facilities required. The same problems must be faced by an individual dairy or a central composting center, but may be more economically dealt with by a larger facility that has its own management, capital, and labour force.

Marketing compost

The main challenge farmers must address before starting on-farm composting or cooperating in the organisation of a central composting center is to determine whether the final product can be marketed successfully and economically. It is better to examine the potential market for the product before beginning production, especially given the small profit margin estimated here. For most dairies, composting would represent a new enterprise. Accurately assessing the potential market often determines the success or failure of a venture. Thus, the farmer should consider how much of the product can be sold and at what quality and price.

References

Falk, Dean E., and Robert M. Ohlensehlen, "Dairy Waste Management System Planning –*Estimating Storage*", *Cooperative Extension, Bulletin No.* 694, University of Idaho, 1989.

Hansen, Brandon D., "An Economic Model for Analyzing Alternative Dairy Waste Handling Systems," M.A. thesis, Department of Agricultural Economics, Washington State University, Pullman, December 1993.

Morgan, Russell M., and Luther H. Keller, "Economic Comparisons of Alternative Waste Management Systems on Tennessee Dairy Farms", *Bulletin 656*, University of Tennessee Agricultural Experiment Station, Knoxville, 1987.

Rynk, Robert, *On-Farm Composting Handbook*, Northeast Regional Agricultural Engineering Service. Ithaca, NY, 1992.

9

Managing Dairy Feeding Programme

Dairy cattle require specific amounts of nutrients to support various levels of performance, so changes in feed intake can have a dramatic impact on the formulation of rations and nutrient intake. Dietary nutrient densities are minimised when feed consumption is maximised, making it easier to formulate rations that are adequate in nutrients. The amount of feed that a dairy cow consumes is highly correlated to its nutrient intake. Every effort should be made to maximise feed consumption when feeding dairy cattle. As feed consumption declines, dietary nutrient densities are increased. The higher the intake, the more forage that can be included in a dairy ration and the fewer concentrates that will be required.

The most cost-effective feeding programmes can be implemented when feed consumption is maximised. Maximised feed consumption minimises the cost of providing required nutrients because higher levels of forages and by-product feeds can be incorporated into the ration. When feed consumption is maximised there is more flexibility in the type of feeds that can be used in formulating the ration.

The quality of forage has a dramatic effect on feed consumption. Feeding the highest-quality forage will maximise feed consumption and nutrient intake and minimise dietary nutrient densities, ration cost and the quantities of concentrates that need to be incorporated into a ration. The feeding of roughages containing high fibre and low digestible energy levels is the primary cause of many dairy farms' failure to realise maximum dry matter intake. Higher forage levels also help to maintain a more stable and healthier rumen and reduce the animal's consumption of grain, which can then be put to other uses, including human consumption.

As forage quality declines, the digestive passage rate becomes slower, resulting in a greater fill factor and causing a reduction in feed intake. When low-quality forages have to be fed, they should be chopped to minimise their depressing effect on feed consumption. Care needs to be taken not to chop forage so fine that milk butterfat is depressed by the resulting low effective fibre level.

Every effort should be taken to minimise heat stress so that feed consumption is not depressed. Shade is very important in areas where cows are exposed to high levels of solar radiation. In hot dry climates, applying water or using misters or evaporative coolers can be effective ways of lowering ambient temperatures, cooling cows and reducing heat stress. Circulating air with fans increases evapotranspiration and increases the dissipation of a cow's body heat load. Opening housing facilities to increase air circulation during times of heat stress increases air movement, which will increase cooling and reduce heat stress. Providing a cool water supply can also help to reduce heat stress.

Adequate consumption of water is critical for maintaining feed consumption; there is a high correlation between feed and water consumption in dairy cattle. When dairy cattle are required to consume poor-quality water, water and feed consumption will be depressed. Maximum performance will only be achieved when cattle have ad libitum access to a good-quality water source.

Poor-quality water or an inadequate supply of water will depress an animal's performance more quickly and more dramatically than any other nutrient deficiency. Adequacy of watering space must also be considered. When cows have to wait too long to drink, their water and dry matter consumption are decreased. If water sources of variable qualities are available, the highest-producing cows should be given the best-quality water.

When developing feeding programmes for dairy cattle, the goal should be to maximise feed consumption by trying to minimise the various factors that depress it. When feed consumption is maximised, performance is optimised, and this should be the primary goal of a dairy feeding programme.

Versatility of Dairy Cattle

Dairy cattle are unique in having the ability to convert cellulytic feed resources (forages, by-products, etc.) that are not suitable for human and monogastric animal (swine, poultry, etc.) consumption into a highly nutritious product, which is milk. Milk is high in nutritive value which can be used directly by the people who own the cows, or be sold as a means of generating income. In some parts of the world, as few as one or two cows can provide a substantial source of daily income for a family for an extended period of time, if adequate feed resources are available and the cows are fed and managed properly.

Dairy cattle are very adaptable since they have the ability to convert a wide range of feed resources into milk. They can be grazed on a wide range of forages (improved and unimproved pastures, roadside forage, forage trees, etc.) or be maintained in partial or full confinement where they can be fed a wide range of harvested feeds (concentrates [such as cereal grains or oilseed meals]; ensiled crops [such as maize, sorghum or barley and various forage haylages/silages]; wet brewers' grains; agricultural processing wastes; various types of hays; and various by-product feeds [such as rice hulls, wheat bran and brewers' grains]); or crop residues (straw, maize stover, etc.). The manure produced by dairy cattle can also be a valuable resource and is used as either a fuel or a fertiliser in various parts of the world.

The versatility of dairy cattle makes them unique among livestock species. They can be maintained in highly productive systems, where feeding and management inputs are very high, and in subsistence-type systems, where inputs are very low. Dairy farming provides a nutritious marketable product without sacrificing the animals.

Developing a Dairy Feeding Programme

The ideal dairy feeding programme is one that optimises the use of available feed resources, so that profitability associated with milk production can be maximised. It has often been thought that one ideal feeding programme could be universally applied all over the world, and an example of such a supposedly universal programme would be the maize-soybean feeding programmes that have been developed and used extensively for swine.

When maize and soybean meal are the most economic feed resources available, the maize-soybean programme would most likely be the feeding programme of choice but, for various reasons, it has often been used where maize and soybean meal are not the most economic feeds available. Dairy cattle do not have a standard feeding programme that can or should be universally applied. In fact, dairy feeding programmes need to be customised for individual farms, and ideally for individual animals, so that they can take advantage of the feed resource that are available to individual producers.

The feed resources available even to neighbours can vary dramatically: one farmer might have pasture to graze, while the neighbouring farmer does not; another farmer might have hay to feed, while the next farm has only straw. In areas or regions where feed resources are available at similar prices, similar feeding programmes can be used, but even then, if the production levels of individual cows vary, different amounts of forage and concentrate will need to be fed.

This means that feeding programmes should be customised to individual producers or regions based on the prices of feeds, the availability of feed resources, the feeds' nutrient content and availability and the milk producing ability of the cows.

Fortunately, some basic nutritional principles can be applied to the development of dairy cattle feeding programmes that allow for a wide range of feed resources to be effectively utilised in the production of milk. The reason dairy cattle are so versatile is that they are ruminant animals and have the ability to convert a wide range of carbohydrate substrates (cellulose, starch, etc.) in their rumen into nutrient sources (volatile fatty acids, microbial proteins, vitamins, etc.) that the cow can then absorb and use to produce milk.

A wide variety of feed resources can therefore be fed to dairy cattle, including forages, crop residues (straws), by-products (rice hulls, wheat bran, beet pulp, etc.), silages and concentrate feeds (cereal grains and oilseed meals). In developing countries where traditional feed resources such as forages (alfalfa, ryegrass, etc.) and concentrates (maize, sorghum, soybean meal, cottonseed meal, etc.) are not readily available or not economically feasible for feeding, a wide range of other feedstuffs can be used to provided the nutrients required to produce milk. The challenge in both developed and developing countries is to optimise milk production while using available feeds to provide the required nutrients in the most economic way possible.

When formulating rations, nutritionists have often aimed at increasing the efficiency of production. Simply stated, they have tried to formulate feeding programmes that would convert the highest amount of nutrients consumed by the dairy cow into milk components (butterfat, protein, etc.). Using this approach they have always tried to maximise production, because the higher the production the more efficiently the cow converts the nutrients that it consumes into milk components.

As an example of efficient dietary nutrient conversion, three feeding programmes were developed for three milk production levels (40, 20 and 10 kg per day) for a healthy 600 kg cow that is maintained under ideal conditions and has the genetic potential to produce 40 kg of milk (containing 3.5 percent butterfat and 3.2 percent milk protein) a day.

There is an improvement of approximately 40 percent in feed nutrient utilisation efficiency between the 40 kg and the 10 kg milk production levels. It is often assumed that maximising a dairy cow's milk production is always desirable (and this would be correct if feed resources were not a limiting factor), and the terms "maximising" and "optimising" milk production are sometimes used interchangeably when talking about formulating feeding programmes for lactating dairy cattle. A ration that maximises milk production is one that maximises the expression of the genetic milk producing ability of the cow, while maintaining the health of the animal's digestive system. However, the feed resources available are often not suitable for maximising milk production because they do not contain the necessary nutrients, contain factors that limit nutrient availability or contain substances that cause nutrient intake to be depressed.

When the economics of milking production are being considered, maximising performance does not always equate to optimising profitability. In many situations, available feed resources are not suitable for maximising milk production. This is especially true in developing countries, where it is often not economically feasible to feed concentrates (cereal grains and supplemental protein sources) to dairy cattle and, therefore, it is not normally possible to provide the adequate levels of dietary nutrients for cows to express their full genetic milk producing ability. Under these conditions it becomes necessary to formulate feeding programmes that will produce milk in the most economic way and optimise milk production (produce the most milk possible from the available resources and at the least cost), but not maximise it.

If feed quality and price are not limiting, the objective should be to maximise production, while maintaining proper digestive tract health, but when feed quality or price are limiting, the goal should be to optimise milk production. Optimising production means producing the most milk for the least cost using the available feed resources.

In most situations, dietary energy and crude protein (CP) cost the least when they are provided by forages, so maximising the amount of forage in a ration, while still providing the other required nutrients, will usually minimise the cost. Typically, least-cost computerised ration formulation systems rank feedstuffs in terms of costs to provide a certain amount (1 kg, 100 kg, etc.) of a specific nutrient (total digestible nutrients, CP, calcium, phosphorus, etc.). Feeds are incorporated into the formulation on the basis of this ranking and starting with the least expensive source of a given nutrient. When the nutrient costs of forages and concentrates are similar, the cost of providing nutrients will not change as much as when feed intake is reduced and the ration needs to be reformulated.

In many developed and developing countries, energy often becomes cheaper when cereal grains are used. If the supply of cereal grains is limited and has to be used for human consumption (as is the case in some developing countries), such price considerations become irrelevant and the use of forages and locally available by-products will need to be maximised in dairy feeding programmes. The higher the feed intake, the more the forage that can be included in a dairy ration and the less the cereal grain that is required.

Regulation of Feed Consumption

The mechanisms involved in feed consumption and appetite regulation are complex. The following factors that are involved in the regulation of feed consumption in dairy cattle.

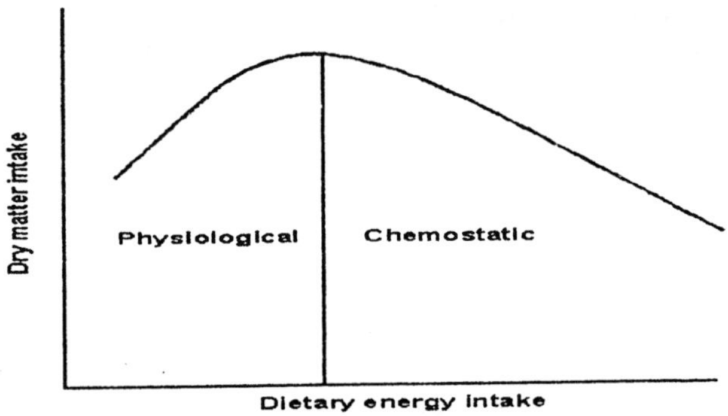

Figure 1. Regulation of feed consumption

The two most important factors for dairy cattle that are not under any type of stress are the physiological and the chemostatic regulatory mechanisms of feed comsumption. The relationship between these two factors is shown in the Figure.

Physiological regulation normally occurs when less digestible feeds, such as low-quality forages or bulky feeds (hulls, etc.) are being fed. Chemostatic regulation occurs when less bulky feeds that contain higher digestible nutrient contents are used.

Physiological Regulation

Physiological regulation is based on the volumetric capacity of the digestive tract, in dairy cattle this specifically relates to the capacity of the rumen. Sensors located in the rumen of the dairy cow sense when the rumen is distended, at which time a signal is sent that causes the animal to stop consuming feed. This type of regulation occurs with feeds that contain low digestible nutrient densities and are bulky, such as straws and other low-quality forages.

Dramatic reductions in consumption occur when these types of feeds are fed. These feeds normally have low digestibilities because they are high in lignin, silica and acid detergent fibre, which depress the digestibility of the nutrients that they contain.

As forage quality declines, the digestive passage rate becomes slower, resulting in a greater fill factor and causing a reduction in feed intake. This same type of reduction in feed intake also occurs when low-bulk density feedstuffs are mixed into complete rations. In most cases, physiological fill is a factor only when dairy cattle are being fed very low-quality feeds.

Chemostatic Regulation

This mechanism functions when blood levels of specific metabolites rise, sending a signal that causes the animal's appetite to be depressed. In the case of dairy cattle, volatile fatty acids are the metabolites that cause the signal to be sent; a few hours after a cow has consumed a meal, the volatile fatty acid levels in the rumen start to rise as a result of rumen fermentation of the ingested substrates.

The dietary digestible energy levels are directly related to the amounts of the metabolites that are produced. Peak volatile fatty acid production normally occurs in the rumen two to three hours after a high-concentrate (high in readily available carbohydrates) ration has been consumed and four to five hours after a high-forage (high in cellulose) ration has been fed. The volatile fatty acids that are produced in the rumen are then absorbed and the levels in the blood rise. Once a certain level of volatile fatty acids in the blood has been reached, the appetite of the animal will be depressed. The volatile fatty acids are continuously absorbed and metabolised by the cells, so when the blood volatile fatty acid level declines the animal's appetite will increase again.

Many other factors have been shown to have an influence on feed consumption, including health, parasite load and digestive disorders. The following sections discuss the effects that forage quality, environmental stress and water quality have on feed consumption in dairy cattle.

Relationship between Nutrient Consumption and Milk Production

The amount of feed that a dairy cow consumes is highly correlated to its nutrient intake. The level of available nutrients determines how much milk a dairy cow is able to produce. The available nutrients can either come from what the cow is consuming in its feed or be taken from its body reserves. If dietary nutrient consumption is not enough to satisfy the nutrient requirements for the animal's level of milk production, the animal will have to mobilise its body nutrient reserves in order to provide the missing nutrients. The nutrient reserves that an animal normally mobilises are energy (fat) and protein (tissue). When this happens, the animal loses body weight, which is normal in high-producing dairy cattle.

The combination of nutrients provided by the diet and derived from body reserves must be sufficient to supply the required nutrients for the amount of milk being produced. If adequate nutrients cannot be derived from these two sources, the cow will reduce its milk production to match the available level of nutrients: nutrient input (diet + body reserves) = output (milk + body composition). When cows are being fed at a high nutrition level but do not have the genetic ability to produce the amount of milk that their feed would allow them to produce, they will deposit the excess energy that they consume as body fat, thus gaining weight. Care therefore should be taken to ensure that the proper amounts of nutrients are provided to support the level of milk production that a cow is genetically capable of producing.

The feed consumption of dairy cattle changes as their productive status changes. They consume different amounts of feed during different stages of their lactation cycles, and different amounts when they are dry and not lactating. For example, dry matter consumption for a 600 kg dry cow that is 40 to 60 days from calving is 9 to 12 kg; when the same cow is ten to 15 days from calving, it will consume 11 to 13 kg of dry matter; rising to 24 to 27 kg when it is producing 45 kg of milk. When the milk production of a dairy cow increases, its feed consumption also increases.

Importance of Forage Quality

Forage is the most important component in the diet of dairy cattle because of the dramatic impact it has on dry matter and nutrient consumption. The quality and form of forage are two of the factors that have been shown to influence dry matter consumption and milk production in dairy cattle.

Forage Quality

Forage quality can be defined simply as the ability of the dairy cow to digest and utilise the nutrient components provided by the forage source. The higher the content and digestibility of the nutrients, the higher the quality of the forage. The highest-quality and most digestible forage is young herbage, because it contains the lowest amount of structural carbohydrates (cellulose, hemicellulose) and lignin.

As a forage matures, its digestibility, rate of digestion and CP content decline, causing the cow to derive fewer nutrients from the forage. A decline in the quality of forage has an impact on the amount of other feedstuffs that the animal is able to consume. The slower passage time of the forage results in a reduction in intake of not only the forage but also other feeds that the animal is consuming.

The quality of a forage declines as it matures. The primary reason for this arises from reduced digestibility, which is related to increases in acid detergent fibre and lignin. Table 1 shows that, as alfalfa matures, its digestibility and CP content decline, reducing the amounts of nutrients that the cow can obtain from the alfalfa and, thus, also reducing intake. Forage quality has also been shown to have an effect on dry matter consumption, especially when low-quality forages are being fed.

Table 1. Effect of maturity of alfalfa on its digestibility

Milk production	Dry matter intake as percentage of body weight for:			
	Digestibility (%)	Crude protein (%)	Acid detergent fibre (%)	Lignin (%)
Prebud	66.8	24	23	4
Bud	65.0	22	25	5
Early bloom	63.1	20	28	6)
Mid-bloom	61.3	19	31	7
Full bloom	59.4	17	33	8
Late bloom	57.5	15	35	9
Mature	55.8	13	38	10

The quality of the forage being fed to dairy cattle has a dramatic impact on not only dry matter consumption but also the proportion of nutrients that are being provided by other feedstuffs. Table 2 gives an example of the effect that changing forage quality has on nutrient consumption.

Table 2. Effects of different qualities of forage on forage and concentrate consumption[1]

Nutrient requirements for producing 20 kg/day of milk		
Total digestible nutrients (TDN) = 10.26 kg		
Crude protein (CP) = 2.086 kg		
Dry matter (DM) intake = 16.2 kg		
Specifications for feedstuffs	TDN	CP
Good-quality alfalfa hay		60 %
18.0 %		
Poor-alfalfa quality hay	50 %	13.0 %
Wheat straw	40 %	3.6 %
Wheat bran	70 %	17.1 %
Good-quality alfalfa hay	10.85 kg	
Wheat bran	5.35 kg	
(CP content	2.87 kg)	
Poor-quality alfalfa hay	5.35 kg	
Wheat bran	10.85 kg	
(CP content	2.55 kg)	
Wheat straw	3.61 kg	
Wheat bran	12.59 kg	
(CP content	2.28 kg)	

[1] Based on the same DM intake and providing enough nutrients for a 600 kg cow to produce 20 kg of milk (3.5 percent butterfat and 3.
[2] percent milk protein).2 All rations were formulated to provide 16.2 kg DM intake and 10.26 kg of TDN.

In the Table 2 example, the amount of concentrate (in this case wheat bran) required to maintain the same energy intake increases from 5.35 kg when good-quality alfalfa is fed, to 12.59 kg when straw is fed. Whenever the quality of the forage declines, the amount of concentrate required to be fed increases, if the same dietary energy level is to be maintained. In Table 2 the amount of dry matter consumption remains constant, but as forage quality declines, dry matter consumption also declines, so even greater quantities of concentrate will have to be fed.

In Table 2, the CP intake ranges from 2.87 kg down to 2.28 kg, which is still above the 2.09 kg minimum required to produce 20 kg of milk. As forage quality declines, the quantity declines and the amount of CP it provides decreases. Lower

CP intakes also have a tendency to reduce feed intake because CP stimulates rumen fermentation, which increases dry matter intake.

Normally, 40 percent of roughage is considered the minimum level required when formulating ratios for lactating dairy cattle. The length of the dietary roughage component must also be considered. An inadequate amount of roughage or reducing the length of the roughage, so that there is not enough effective fibre, will cause butterfat depression and, often, digestive problems.

Visual Appraisal of Forage Quality

Visual appraisal of forage can be useful in assessing its quality. The maturity of a forage can be estimated quite accurately by the number of buds, blossoms or seed heads that are present. Proper curing during the haymaking process can be assessed by the colour of the hay. Colour can also be used to assess the extent of nutrient losses associated with leaching resulting from exposure to rain and weather.

Bleached forages will have lower vitamin and CP contents. With legume-type forages, the leaf-stem ratio can provide a fairly accurate estimate of the nutrient content of the forage. When there are many leaves, the CP content is high; when there is more stem, the structural carbohydrates content will be higher and the digestible nutrients content lower. The CP content of a forage is closely correlated to its digestibility - the higher the CP content of a forage, the higher its digestibility will be.

Palatability of Forage

The palatability of a forage is affected by its taste (sweet, salty, bitter, acidic), olfactory and textural characteristics. Taste is normally the major factor affecting palatability. Dairy cattle are non-selective consumers and readily consume a wide range of feeds. Almost all livestock show a preference for sweet, so feed consumption can often be increased by adding molasses to a ration. Salt can also be used to increase the palatability of a feed but, once it reaches a certain level, increasing the salt content will depress feed consumption. Palatability can play a role in feed consumption when the animals have a choice, but dairy cattle do not usually have a choice, so palatability is not a major factor in feeding dairy cattle. Palatability normally becomes a factor only when attempts are made to feed spoiled feeds to dairy cattle.

Processing of Forage

The decrease in intake that occurs as a forage matures can be counteracted to some extent by reducing the physical size of the forage (through chopping or grinding), which will allow it to pass through the rumen at a faster rate. The passage rate out of the rumen is based on particle size and density. Small, dense particles are passed out of the rumen more quickly than larger forage particles (most of which are less dense and float), which are retained in the rumen.

As the ruminal passage rate increases, exposure to the digestive processes decreases and the overall digestibility of the forage declines but, because more can pass through the digestive tract, the animal will increase its dry matter consumption and the net result is usually that the cow's digestible nutrient intake increases slightly. This is one of the reasons for chopping forages prior to feeding. Chopping is most beneficial when low-quality forages are fed, but forages should not be chopped into pieces that are too small, as this can result in a depression of milk butterfat.

Influences of Heat Stress on Feed Consumption

Heat stress is another factor that has been shown to have a major impact on the feed consumption of a dairy cow. Feed consumption decreases during hot weather and increases during cold weather. Several other factors have been found to be associated with, and have an influence on, heat stress, including humidity, air movement, shade and availability of water.

Temperature and Humidity

Factors that influence the body temperature of dairy cattle can cause stress and have been shown to have dramatic effects on feed consumption. Both low and high temperatures can cause temperature stress and both can have impacts on nutrient consumption and nutrient utilisation. As the core body temperature of an animal increases, the hypothalamus causes the cow's appetite to be depressed. This results in a depression of dry matter intake and can have a dramatic impact on nutrient intake and milk production. The effect that temperature has on feeding consumption is shown in Table 3.

Table 3. Intake adjustments for different environmental temperatures

Temperature	Intake adjustment (%)
> 35°C, no night cooling	- 35
> 35°C, with night cooling	- 10
25 to 35°C	- 10
15 to 25°C	None
5 to 15°C	3
-5 to 5°C	5
-15 to -5°C	7
< -15°C	16

Heat stress has a more dramatic impact than cool stress on feed consumption and milk production. It is not only related to ambient temperature, but also associated with humidity and air movement. When the humidity increases, the cow's evapotranspiration is reduced and the animal cannot cool itself, which increases its core body temperature and depresses feed intake. Temperature alone is not a good way of measuring heat stress, so various heat indexes have been developed which take into account such factors as temperature, humidity and evaporation rate. The rectal temperature of dairy cattle has been found to be one of the best indicators of heat stress.

The duration of heat stress also influences feed consumption. High daytime temperatures can be tolerated if cooling occurs at night and the cows are able to dissipate the body heat that has built up during the day. The most severe heat stress occurs when both humidity and temperature are high and the night-time temperature does not decrease, so the cows cannot dissipate their body heat. In addition to depressing feed consumption, heat stress has also been shown to have an effect on milk composition. Milk protein percentages have been shown to decrease during periods of heat stress, and some reduction in milk butterfat has also be observed in dairy cattle.

Ways to Minimise Heat Stress

During periods of heat stress cows will consume more water, so an adequate supply of water needs to be provided. Providing a cool water supply can help to reduce heat stress. Cows will consume more cool water during periods of heat stress; although it is not the temperature of the water but rather the additional evapotranspiration that helps the cows to cool. Providing adequate shade will help reduce the cows' uptake of solar heat, and shade is particularly important in areas where cows are exposed to high levels of solar radiation.

Cooling the cows with misters and evaporative coolers can also be an effective means of lowering ambient temperatures and reducing heat stress in hot, dry climates. Circulating air with fans will increase evapotranspiration and increase the dissipation of a cow's body heat load. Opening housing facilities during times of heat stress will increase air movement which will increase cooling and reduce heat stress.

Water Quality

Water is one of the most important nutrients that an animal consumes. Numerous important biological processes require water, such as digestion, absorption, transport and excretion of nutrients and metabolites, components of milk, body temperature regulation and cellular metabolism. Death occurs about nine to ten times more quickly as a result of water deprivation than because of feed deprivation. Poor-quality water or the lack of an adequate supply will depresses an animal's performance more quickly and more dramatically than any other nutrient deficiency.

The amount of available water and the water quality are often overlooked when developing feeding programmes for dairy cattle. When the amount or quality of water becomes restrictive, an animal will not perform at the maximum of its genetic potential. If cows are only able to drink once or twice a day, they will produce less milk; and if adequate water is not available or the quality of the water is low, feed consumption will be reduced and performance will be depressed. Dry matter intake has been found to be highly correlated (at a ratio of 0.91) with water consumption.

Factors Influencing Water Intake

Water requirements vary considerably and are related to the type of diet being consumed and the environmental conditions under which the animal is being maintained. At 4°C, a 545 kg non-lactating cow requires 30 kg/day of water, but at 32°C the same cow will require 57 to 72 kg of water, depending on the humidity, because of increases in body water losses. The water content of the ration being fed can also have a dramatic effect on the amount of drinking-water that is required.

Table 4. Relationship between environmental temperature and water requirements of livestock

Environmental temperature (kg/kg DM consumed)	Water requirements
> 35°C	8 to 15 kg
25 to 35°C	4 to 10 kg
15 to 25°C	3 to 5 kg
-5 to 15°C	2 to 4 kg water
< -5°C	2 to 3 kg[1]

[1]Increases of 50 to 100 percent occur when there is a rise in environmental temperature following a period of very cold temperature, e.g. a rise from -20 to 0°C

For example, 20 kg of maize silage will provide 13.4 litres of water. In areas where water quality or supply is low, the use of feeds that contain water, such as wet brewers' grains or silage, can be an effective way of increasing milk production. Environmental conditions such as temperature, humidity, wind movement and exposure to sun all alter the water requirements of an animal, as shown in Table 4.

Performance will only be maximised when good-quality water is freely available. It is important to have an adequate supply of acceptable-quality water available to the animals at all times. Non-lactating cattle consume approximately two-and-a-half to three times as much water as dry matter, and lactating cattle consume 4.5 to 5 kg of water for each additional kilogram of milk produced. When water intake is reduced, dry matter intake will also be reduced, resulting in reduced milk production. Diets that are high in CP or salt increase water requirements further.

Water intake in lactating dairy cattle is influenced by milk production, ambient temperature, humidity, salt intake, dry matter intake and other factors. Elevated environmental temperatures increase the consumption of water because of the increased losses that the cow undergoes as a result of increased evapotranspiration. More water should therefore be provided during periods of elevated temperatures, or feed intake will be depressed. If water quality is marginal, the increase in water consumption can become problematic and consuming more water can put additional physiological stresses on the animals. Animals that do not have access to an adequate supply of water will consume less feed and produce less milk.

Factors Influencing Water Quality

The following factors can effect the quality of the water: alkalinity, total dissolved solids, specific minerals (nitrates, sulphates, etc.), and bacterial or algae content. Alkalinity is a measure of how much acid is required to neutralise the pH of a water supply. "Total dissolved solids" refers to the dissolved inorganic salts that are present in the water supply.

Nitrates can be found in water supplies at levels that are toxic to dairy cattle because they are converted to nitrites in the rumen. Sulphates can cause diarrhoea, which will reduce the efficiency of nutrient absorption.

Table 5. Guidelines for total dissolved solids in water for livestock and poultry

Total dissolved solids (ppm)	Quality of water source
< 1,000	Excellent water source that can be used with all classes and types of livestock.
1,000 to 2,999	Water should be acceptable for all classes and types of livestock. Water approaching the upper limits may cause watery droppings in poultry, but performance should not be affected.
3,000 to 4,999	Water should be satisfactory to marginal for most livestock. Animals that are not used to drinking it might take a few days to adjust. If sulphate levels are high, animals might have diarrhoea. Performance will be only slightly depressed. Poor source of water for poultry, which would have increases in faecal water output; at the upper limits this water decreases growth and increases mortality, especially in turkeys.
5,000 to 6,999	Can be used for livestock, but some depression in productivity and physiological condition is likely to occur. There will be higher refusal rates and the water is likely to have a laxative affect. Should not be used for poultry.
7,000 to 10,000	Poor-quality water that will affect performance. Can be used for mature ruminant animals that are just being maintained. Should not be used with pregnant or lactating animals. Is not suitable for poultry or swine.
> 10,000	Unsatisfactory for all types and classes of livestock.

A water supply that is too alkaline can cause physiological and digestive problems. Bacterial and algae contamination do not usually affect performance, except when they cause water palatability problems which lead to reduced water consumption. Some algae contain compounds that are toxic to animals if consumed in large quantities. Tables 5 and 6 show the recommended guidelines for water for use with livestock and poultry.

Table 6. Water quality guidelines for livestock

Mineral element	Maximum tolerances (mg/litre)
Macrominerals	
Calcium	1,000
Nitrate and nitrite	100
Nitrite only	10
Sulphate	1,000
Total dissolved solids	3,000
Heavy and microminerals	
Aluminium	5
Arsenic	0.5
Boron	5
Cadmium	0.02
Chromium	1
Cobalt	1
Copper	5
Fluoride	2
Lead	0.1
Mercury	0.003
Molybdenum	0.5
Nickel	1.0
Selenium	0.05
Uranium	0.2
Vanadium	0.1
Zinc	50

The most common minerals present in water include chlorine (Cl), sodium (Na), calcium (Ca), magnesium (Mg), sulphate (SO_4) and bicarbonate (HCO_3). The specific minerals that are present in the water depend on soil type and the source of the water. The tolerance of animals to alkaline (dissolved salt) water depends on several factors such as water intake, species, age, physiological condition, season of the year and salt content of the diet.

Several factors should be considered when assessing the quality of an available water source. Generally, water containing more than 1 percent NaCl (common salt) is not considered good quality because this is about the maximum salt content that cattle and sheep can tolerate without decreasing their productivity. If there is a question about the quality of a water source then the source should be tested. Mineral content is important, especially in arid areas.

Depending on the type of minerals that are present, up to 15,000 mg/litre (1.5 percent) of dissolved solids may be tolerated by livestock, but normally the palatability of the water is reduced as the mineral content increases and the animal's performance will decline; a good-quality water source should therefore contain less than 2 500 mg/litre (0.25 percent) of dissolved solids. Some salts, such as nitrates, fluorine and other heavy metals, may become toxic before the levels that affect palatability are reached. Levels of 100 to 200 parts per million (ppm) of nitrates are potentially toxic and 1 g of sulphate per litre may result in diarrhoea. Other materials that are sometimes found in water supplies and may affect palatability or be toxic include pathogenic microorganisms, algae and protozoa, hydrocarbons, pesticides and many industrial chemicals.

Ways to Improve Water Quality

Water can be derived from various sources - wells, ponds, rivers/streams, springs, etc. The most commonly used source of water for livestock is surface water (streams, ponds, etc.). Rainwater that has been collected and stored can also be a good source of water in some areas. If water sources of different qualities are available, then the highest-producing cows should be given the best-quality water.

The adequacy of watering space must also be considered. If cows have to wait too long to drink, their water consumption and dry matter consumption will be decreased.

The drinking-water source should be located near where the animals are being fed; if animals have to go long distances to water they will consume less. In order to achieve maximum performance, water sources need to be available in close proximity to grazing areas or the areas where cows are being housed. Cattle can be prevented from walking and defecating in the water by piping it into watering tanks. Maximum performance will only be achieved when cattle have ad libitum access to a good-quality water source.

References

Akin, D.E. & Chesson, A., "Lignification as the major factor limiting forage feeding value especially in warm conditions", *Proceedings of the XVI International Grassland Congress*, Nice, France, 1989.

Avendano, L., "Reduction of heat stress in dairy cattle through the utilization of cooling systems", *Cuban Journal of Agricultural Science*, 1995.

Bernabucci, U. & Calamari, L., "Effect of heat stress on bovine milk yield and composition", *Proceedings of the Milk Yield and Quality in Cows Raised in Hot Environment Seminar*, Zootecnica e Nutrizione Animale, 1998.

Cecchini, G., "Keep it cool for cows", *African Farming and Food Processing*, 1998.

Chalong-Wichiraphakorn, "Methods of reducing heat stress in dairy cows", *Khon Kaen Agriculture Journal*, 1995.

Dado, R.G. & Allen, M.S., "Variation in and relationship among feeding, chewing and drinking variables for lactating dairy cows", *Journal of Dairy Science*, 1994.

Fahey, G.C. Jr. & Hussein, H.S. "Forty years of forage quality research: accomplishments and impact from an animal nutrition perspective", *Crop Science*, 1999.

10

Hygiene Management in Dairy Farms

The milk secreted into an uninfected cow's udder is sterile. Invariably it becomes contaminated during milking, cooling and storage, and milk is an excellent medium for bacteria, yeasts and moulds that are the common contaminants. Their rapid growth, particularly at high ambient temperatures can cause marked deterioration, spoiling the milk for liquid consumption or manufacture into dairy products. This can be avoided by adopting the simple, basic rules of clean milk production.

Udder Infection

The essential requirements are to maintain udders free from infection (eg. mastitis); manage cows so that their udders and teats are clean; milk them in such a way that minimises bacterial contamination; store the milk in clean containers and, wherever possible, at temperatures which discourage bacterial growth until collected. Simple and low-cost husbandry practises enable milk to be produced with a bacterial count of less than 50,000 per ml. The golden rule of clean milk production is that prevention is better than cure.

It is impossible to prevent mastitis infection entirely but by adopting practical routines it can be kept at low levels. Most mastitis is subclinical and although not readily detected by the stockman, it will not normally raise the bacterial count of herd milk above 50,000 per ml. Once the clinical stage is reached, the count may increase to several millions/ml and one infected quarter may result in the milk from the whole herd being unacceptable. It is important to detect clinical cases and exclude their milk from the bulk.

Under normal grazing conditions, cows' udders will appear clean and therefore washing and drying will be unnecessary. Otherwise, any visible dirt must be removed using clean, running water, individual paper towels or cloths in clean water to which a disinfectant has been added (eg. sodium hypochlorite at 300 ppm). If udder cloths are used, provide a clean cloth for each cow. After each milking wash and disinfect them and hang up to dry. Disposable paper towels are preferable and more effective for drying after washing. When cows are housed or graze in heavily stocked paddocks, external udder surfaces are usually grossly contaminated with bacteria even when they appear visibly clean, therefore routine udder preparation procedures should be followed. Whenever udders are washed they should be dried.

Foremilking has little affect on the total bacterial count of the milk but is an effective way of detecting clinical symptoms of mastitis. Filtering or straining the milk removes visible dirt but not the bacteria in the milk because they pass through the filter. Aerial contamination of milk by bacteria is insignificant under normal production conditions.

The milk contact surfaces of milking and cooling equipment are a main source of milk contamination and frequently the principal cause of consistently high bacterial counts. Simple, inexpensive cleaning and disinfecting routines can virtually eliminate this source of contamination.

Cleaning Milk Production Equipment

It is virtually impossible with practical cleaning systems to remove all milk residues and deposits from the milk contact surfaces of milking equipment. Except in very cold, dry weather, bacteria will multiply on these surfaces during the interval between milkings, so that high numbers (eg 106 per m²) can be present on visually clean equipment. A proven cleaning and disinfectant routine is required so that with the minimum of effort and expense, the equipment will have low bacterial counts as well as being visually clean.

The essential requirements are, to use milking equipment with smooth milk contact surfaces with minimal joints and crevices, an uncontaminated water supply, detergents to remove deposits and milk residues and a method of disinfection to kill bacteria.

Water Supplies

Unless an approved piped supply is available it must be assumed that water is contaminated and therefore hypochlorite must be added at the rate of 50 ppm to the cleaning water. Hard water (ie. high levels of dissolved calcium and other salts) will cause surface deposits on equipment and reduce cleaning effectiveness. In such cases, it is necessary to use de-scaling acids such as sulphamic or phosphoric, periodically.

Detergents and Disinfectants

Detergents increase the 'wetting' potential over the surfaces to be cleaned, displace milk deposits, dissolve milk protein, emulsify the fat and aid the removal of dirt. Detergent effectiveness is usually increased with increasing water temperature, and by using the correct concentration and time of application. Detergents contain inorganic alkalis (eg. sodium carbonate and silicates and tri-sodium phosphate), surface-active agents (or wetting agents), sequestering (water-softening) agents (eg. polyphosphates) and acids for de-scaling. Many proprietary, purpose-made detergents are usually available, but otherwise, an inexpensive mixture can be made to give a concentration in solution of 0.25% sodium carbonate (washing soda) and 0.05% polyphosphate (Calgon).

Disinfectants are required to destroy the bacteria remaining and subsequently multiplying on the cleaned surfaces. The alternatives are either heat applied as hot water or chemicals. Heat penetrates deposits and crevices and kills bacteria, providing that correct temperatures are maintained during the process of disinfection. The effectiveness of chemicals is increased with temperature but even so, they do not have the same penetration potential as heat and they will not effectively disinfect milk contact surfaces which are difficult to clean.

When hot water alone is used, it is best to begin the routine with water at not less than 85°C, so that a temperature of at least 77°C can be maintained for at least 2 minutes. Many chemicals are suitable as disinfectants, some of them combined with detergents (ie. detergent-sterilisers). Use only those which are approved, avoiding particularly those which can taint milk (eg. phenolic disinfectants). Always follow the manufacturers instructions. Sodium hypochlorite is an inexpensive example of an approved disinfectant suitable for most dairy purposes. Sodium hydroxide (caustic soda) can also be very effective at concentrations of 3%-5% at ambient temperatures, providing adequate contact time is given with the surfaces to be cleaned and disinfected.

Dairy disinfectants are sold as concentrates and in this form are often corrosive and damaging to the skin and eyes. They should always be so labelled, handled with care and stored out of reach of children. Disinfectants should not be mixed unless specific instructions are given and disinfectant powders must be kept dry. If any concentrated detergent and/ or disinfectant comes in contact with the skin or eyes the affected area should be washed immediately with copious amounts of clean water. If acids are used they must always be added to the water not vice versa.

Milking Premises

The milking premises should have a dairy or suitable place equipped with a piped hot and cold water supply, a washtrough, brushes, a work surface, storage racks and cupboards and, if necessary, a vacuum pipeline connection. In addition, it is advisable to have a dairy thermometer (0°C -

100°C), rubber gloves and goggles for use when handling chemicals.

Daily Routines

Daily routines for cleaning and disinfecting vary with the size and complexity of the milking installation but will include methods of removing dirt and milk from the equipment followed by disinfection. For hand milking, bucket and direct-to-can milking machines, basic manual methods of cleaning and sterilizing are adequate and effective. For pipeline milking machines in-situ (in-place) systems are necessary.

Milk can become grossly contaminated from bacteria on ancillary equipment which must also be cleaned and disinfected effectively. Coolers, either the corrugated surface or the turbine in-can, can best be cleaned and disinfected manually and stored in the dairy to drain. Refrigerated bulk milk tanks can be cleaned either manually using cold or warm detergent/disinfectant solutions, or for the larger tanks, by automatic, programmed equipment. In either case, a cold water chlorinated (50 ppm) rinse preceeds and follows the washing solution. Foremilk cups can be a potent source of bacterial contamination and need to be cleaned and disinfected after each milking. They should then be stored in the dairy to drain.

It is important with any method of cleaning that the equipment is drained as soon as possible after washing for storage between milkings. Bacteria will not multiply in dry conditions but water lodged in milking equipment will, in suitable temperatures, provide conditions for massive bacterial multiplication. Equipment with poor milk contact surfaces, crevices and large number of joints, remaining wet between milkings in ambient temperatures above 20°C, should receive a disinfectant rinse (50 ppm available chlorine) before milking begins.

Mastitis Control

Mastitis is an inflammation of the udder and is common in

dairy herds causing important economic losses. It cannot be eradicated but can be reduced to low levels by good management of dairy cows.

Of the several causes of mastitis only microbial infection is important. Although bacteria, fungi, yeasts and possibly virus can cause udder infection the main agents are bacteria. The most common pathogens are Staphylococcus aureus, Streptococcus agalactiae, Str. dysgalactiae, Str. uberis and Escherichia coli though other pathogens can cause occasional herd outbreaks. Mastitis occurs when the teats of cows are exposed to pathogens which penetrate the teat duct and establish an infection in one or more quarters within the udder. The course of an infection varies, most commonly it persists for weeks or months in a mild form which is not detected by the stockman (ie. subclinical mastitis).

With some pathogens, typically E coli, the infection is frequently more acute and there is a general endotoxaemia with raised body temperature, loss of appetite and the cow may die unless supportive therapy is given. When clinical mastitis occurs the effective therapy is a course of antibiotic infusions through the teat duct. These nearly always remedy the clinical disease and often eliminate the bacterial infection. Infections may spontaneously recover but most persist to be eliminated eventually by antibiotic therapy or when the cow is culled. The susceptibility of cows varies considerably and new infections are most common in older cows in early lactation, at the start of the dry period and when the management is poor.

Mastitis causes direct economic losses to farmers in several ways. Milk yields are reduced, milk that is abnormal or contaminated with antibiotics is unsaleable, there are veterinary and antibiotic costs, a higher culling rate and occasional fatalities. The milk processing industry also incurs losses because of problems that result from antibiotic in milk, and the reduced chemical and bacterial quality of mastitic milk.

Infectious Mastitis

Mastitis microorganisms, usually bacteria, originate in various sites on the cow. They multiply in various ways and are spread from cow to cow. Most common types of mastitis bacteria originate in the udders of infected cows and in sores on tests. These pathogens multiply in teat sores and are spread during milking.

The several microbial diseases of the udder that are collectively known as mastitis are distinctly different. The pathogens can arise from different primary sites, they multiply in different environments and therefore the timing of the exposure of the cow to the bacteria will vary. Subsequently the acuteness and persistency of the infections differ and also the probability of cure when therapy is given.

The commonest forms of mastitis in most countries are caused by S. aureus and Str. agalactiae. The primary sites of these is the milk of infected quarters and therefore they are spread mainly at milking, either during udder preparation or on hands and milking machines. These pathogens can colonise and multiply in teat sores and in teat ducts and this greatly increases the degree of exposure of the teats to bacteria. They usually cause chronic infections which persist in the subclinical form and occasionally become clinical when abnormal milk can be detected. Systemic infection with loss of appetite and raised body temperature is infrequent. When suitable antibiotic preparations are infused into the udder the clinical mastitis nearly always subsides and most Str. agalactiae infections are cured but with staphylococcal infections the cure rate is poor and most persist.

Infections caused by Str. uberis and E. coli are often called 'environmental'. The main primary sites of the pathogens are bovine, but not from within the udder. These do not normally colonise teat skin and the multiplication occurs in organic bedding materials (eg. straws and sawdust). These types of infection are most common in housed cattle in early lactation and whilst they can cause persisting subclinical mastitis the more typical from is clinical mastitis soon after the onset of the infection, and with coliform mastitis the endotoxaemia causes raised body temperature and marked reductions in milk production. Str. uberis infections usually respond to therapy but with E. coli infections it is important to give supportive treatment to overcome the endotoxaemia and if this is successful spontaneous recovery usually follows.

Str. dysgalactiae is similar to Str. agalactiae and S. aureus in that it can readily colonise and multiply in teat lesions but the main primary site is not the milk of infected quarters, but other bovine sites. The course of the infection is not dissimilar to Str. agalactiae and infections respond readily to antibiotic therapy.

Many other microorganisms can cause mastitis. These less common forms are not usually important but pseudomonads and Mycoplasma bovis does cause serious problems in a few herds. Although the pathology of the various types of infection show distinct differences the causes of infection can be diagnosed with certainty only by bacteriological tests made on aseptically taken quarter milk samples.

Causes of Mastitis

For an intramammary infection to occur it is necessary for:

- the teat skin to be contaminated with pathogens
- the pathogens to penetrate the teat duct
- the infection to be established in the sinuses, ducts or tissues of the udder

Exposure to Pathogens

All dairy cows are continously exposed to pathogens that can cause mastitis but new infection is normally infrequent. This is because the exposure to pathogens is usually small. The number of pathogens in the milk of infected quarters will vary from less than 1000 to many millions per ml of milk but it is usually less than 10,000 per ml and further diluted by the milk from the majority of uninfected quarters. The number of pathogens on clean pasture will be extremely small. The exposure of cows teats to S. aureus, Str. agalactiae and Str.

dysgalactiae will be greatly increased when teat sores and lesions on teat ducts are colonised by these mastitis pathogens or when cows lie on contaminated bedding or corrals. Occasionally the exposure will be increased from improperly cleaned milking equipment or through udder washing with contaminated water. This can be avoided by adopting simple methods of cleaning equipment.

The most effective way of reducing the exposure to S. aureus, Str. agalactiae and Str. dysgalactiae is by dipping or spraying the teats in disinfectant immediately after milking. A number of disinfectant products are available and good results are obtained with hypochlorite, iodophor and chlorhexidine. Teat disinfection greatly reduces residual contamination and more importantly encourages healing of teat sores and lesions and also prevents the growth of pathogens in teat ducts (ie. colonisation). Other practices such as udder washing with disinfectant, rinsing milking equipment after each cow is milked (eg. back flushing) will also reduce exposure but their effects are smaller than teat dipping because they do not influence the colonisation of teat sores and ducts. To maintain low levels of exposure it is most important to maintain healthy teat skin and avoid sores, chaps and any form of teat damage. To maintain good teat condition emolients or salves (eg. glycerol) can be added to most teat dips but not to hypochlorite.

While soil and fresh bedding materials are usually relatively free from mastitis pathogens they can develop a very large pathogen population (>106/gm) within a few days given the optimum conditions of moisture and temperature. Inevitably bedding becomes moist and contaminated with faeces and given sufficient warmth the growth of E. coli and Str. uberis is rapid causing outbreaks of mastitis. This can be prevented by avoiding using muddy corrals and by keeping cows on clean pasture. When cattle are housed on straw or sawdust in yards or in cubicles (freestalls), it should be replenished frequently, preferably daily. Straw yards should be cleaned completely if outbreaks of clinical mastitis occur. For cows kept in cubicles sand can be used successfully as a non-organic bedding material in which coliforms and Str. uberis do not readily multiply.

Penetration of the Teat Duct

The healthy teat duct (streak canal) is extremely effective in preventing the passage of pathogens into the udder and it is the cows most natural barrier against mastitis infections. It not only acts as a physical barrier against penetration but the lining of the duct also contains secretions that inhibit bacterial growth. Pathogen penetration can occur during milking, in the intervals between milking and even when cows are not lactating. If pathogens penetrate during milking they may be flushed out otherwise infection usually occurs. It is most important to recognise that even when very small numbers of pathogens (eg. less than 20) penetrate the duct infection usually occurs. Apart from the increased rate of penetration that occurs with high rates of exposure there are other factors in cattle management that are important. These occur with faulty machine milking and also if milkers insert contaminated objects (eg. dirty treatment syringes) through the teat duct.

Under certain milking machine conditions there are considerable vacuum fluctuations in the teat cup liner below the teat. In the resulting turbulence the milk which is normally flowing away from the teats returns to the teat through the short milk tubes (ie. reverse flow or impacts). These impacts can have sufficient force that the pathogens carried in the milk penetrate into or through the teat duct causing infection. These infections can be reduced by using a milking machine fitted with liners that do not slip during milking, that have adequate diameter short milk tubes and an effective airbleed in the clawpiece. Even better protection is achieved by using liners fitted with 'shields' or milking with 'non-return valves' in the milking machine cluster.

Inserting instruments through the teat duct should be avoided but when necessary (eg. for therapy) the teat end should be scrubbed with disinfectant for 15 seconds before insertion and the instrument must be sterile. If in doubt about sterility the instrument should be placed in boiling water for at least 30 seconds, but if this is impractical it should be scrubbed in surgical spirit (70% ethanol).

Establishment of Infection

A number of antimicrobial systems occur in the bovine udder and milk but they appear to be relatively ineffective in preventing infections though they are important in reducing the severity of infections and to a lesser degree in eliminating infections. The hosts defences are partly cellular (leucocytes) and immune defences and in addition there are non-specific defences provided by complex biochemical systems including enzymes and other constituents of milk. Although most types of infection tend to persist for months if untreated the cellular defences will eliminate most E. coli infections providing the cows cellular response to infection is rapid.

This brief description of the prime importance of pathogen exposure, duct penetration and the establishment of infection deals only with the main aspects that are important in considering mastitis control. The interactions are more complex and not fully understood. There are other physiological and environmental factors which are or appear to be important in particular herds. These include nutrition, heat stress and factors such as age of cow and stage of lactation.

Elimination of Infection

Apart from E. coli infections most persist if untreated until eventually there is spontaneous recovery or the cow is culled. This persistance of infection for weeks, months and years with staphylococcal infections is an important characteristic of mastitis which must be taken into account in devising a disease control.

Because spontaneous recovery occurs infrequently and there is no proven way of increasing it the introduction of antibiotic therapy was a major advance in controlling bovine mastitis. Infusing antibiotics via the teat duct into the udder is a simple way of overcoming nearly all clinical mastitis and of eliminating many infections. Many antimicrobial drugs are used in formulating products for mastitis therapy (eg. penicillin including the semisynthetics, streptomycin, aureomycin).

The effectiveness of the drug will depend partly on the sensitivity of the pathogen treated to the drug used and also to the way the drug is formulated. This influences the absorption, distribution, metabolism and excretion of the drugs from the milk. No antimicrobial drug is ideal for all conditions and mixtures are used to be effective against a range of pathogens (ie. 'broad spectrum') for use in lactating and dry cows. Nearly all therapy is given without prior knowledge of the causative pathogen. Under these conditions treating quarters exhibiting clinical mastitis with basic standard products gives a bacteriological elimination of 75-90% of streptococcal and about 30% of staphylococcal infections. The cure rates of mycoplasma and pseudomonad infections are lower. If the treatments are given after the final milking of lactation (ie. dry cow therapy) using formulations designed for this purpose the cure rates are improved and are about 50% for staphylococcal infections. The rates of elimination of staphylococcal infections varies considerably between herds and are lower in cows with more than one infected quarter and those with more severe clinical mastitis.

Antibiotic therapy is usually given as infusions of solutions through the teat duct but this should be done only after careful cleaning of the teat orifice with a disinfectant swab. To avoid infusing contaminated antibiotic, use only factory made disposable syringes. A course of therapy is often 2 or 3 infusions at 24 hour intervals, rejecting the milk from the cow for 2 or more days after the last infusion to avoid antibiotic contamination of the bulk milk. Although nearly all cases of clinical mastitis respond quickly it may take several days before the milk becomes normal. Treatment, the choice of antibiotic and the milk reject time should be on veterinary advice and when there are severe persistent infections systemic administration of antibiotic may also be used to overcome the endotoxaemia.

Many infections are eliminated by culling cows and this can be increased by the sale of cows with persistant clinical mastitis.

Principles of Mastitis Control

Mastitis cannot be eradicated but can be reduced to low levels by adopting simple economic management routines that relate to the patterns of infection. Currently it is not possible to base a control on vaccination and although cows susceptibilities are largely inherent there are major practical limitations to control through breeding, and progress would be slow. Therapy is invaluable to contain the problem but cannot be the basis of a control which must depend on preventing new infections. In practice the key to control is good cattle management particularly steps to reduce exposure to pathogens and also the planned use of antibiotic therapy. Because control depends on management the steps must be simple and economic and fit easily into a milking routine.

There are three main patterns of infectious mastitis.

- The first is the most common and is caused by Streptococcus agalactiae, Staph. aureus and Str. dysgalactiae. These infections usually persist as subclinical infections and this persistance is important. Even if the rate of new infection is considerably reduced the proportion of infected quarters will decline slowly over several years, unless the duration of infection is also reduced. In addition to general good management including sound milking methods, teat disinfection (ie. dipping or spraying) is particularly useful to prevent these infections, by reducing exposure to pathogens. To reduce the persistance of infection the key step is dry cow therapy.
- The second type of infection occurs mainly in housed or closely corralled cattle and is typified by acute clinical mastitis in early lactation. The main pathogens

are Str. uberis and E. coli and their epidemiology is such that they are not controlled by teat disinfection. Drying off therapy is useful in preventing Str. uberis infections that commonly occur early in the dry period but it will not reduce the Str. uberis and E. coliform infections occurring at, and soon after, calving. Control depends primarily on reducing the exposure to pathogens by moving cattle regularly to clean corrals or providing clean bedding, preferably daily.

— The third type of infection occurs in non-lactating cows These are common in the early part of the dry period particularly with Str. uberis and most persist causing clinical mastitis in the following lactation. Another form common in N. Europe occurs later in the dry period and is caused by a complex infection with Corynebacterium pyogenes, Peptococcus indolicus and streptococci. This is 'summer' or 'heifer' mastitis and its aetiology appears to require exposure to pathogens carried by a species of fly. Control is achieved by drying off therapy and fly control.

It is important to recognise that because most mastitis is subclinical and unseen control depends primarily on adopting sound management routines for the whole herd. It cannot be achieved by using laboratory tests to identify individual infected cows and taking special action with these animals. Tests are useful to alert farmers to the extent of the problem but they rarely indicate steps additional to those that should be in the daily routine.

In a concise monograph it is impossible to cover the causes of all types of mastitis. Infrequently a herd mastitis outbreak will occur with an unusual pathogen even when the farmer adopts good control routines. The solution of this type of problem does require investigation by a veterinary diagnostic laboratory in order to discover the source of pathogen and the aetiology.

Control Methods

The following routine will reduce the proportion of infected cows and clinical mastitis by at least 70% if used regularly at each milking. Mastitis caused by Str. agalactiae will be reduced to very low levels and is frequently eradicated.

1. Adopt good cow management practices as the essential basis for a mastitis control routine (eg. feeding, housing, hygiene). Mastitis is unlikely to be controlled with neglected, underfed cows kept under stress in dirty conditions.
2. Reduce exposure to pathogens
   — Clean thoroughly all equipment used when milking.
   — Do not house or corral cattle under dirty conditions, preferably change organic bedding materials daily or use sand for bedding
   — Wash dirty udders before milking with clean running water preferably with the hand, a disposable paper towel or a disinfected cloth and dry thoroughly. Do not wash with contaminated cloths and water
   — Dip or spray all teats after milking with disinectant teat dip (eg. hypochlorite, iodophor, chlorhexidine)
   — Adopt practices that prevent the occurrence of teat lesions (sores, chaps and teat damage). If they occur use a teat dip or spray containing a emolient.
   — If practical milk clinically affected cows last
   — Additional benefits can be obtained by disinfecting hands before milking each cow, using individual paper udder cloths, dipping teat cups in disinfectant before each cow is milked, and 'back flushing'. These reduce bacterial exposure but their effects are secondary to those described above
3. Reduce the chances of pathogens penetrating the teat duct by
   — avoiding teat injury or fly attack
   — using a milking machine that is correctly tested, and maintained

- — using a milking machine modified to prevent 'reverse flow' and 'impacts'
- — minimise the effects of vacuum fluctuations by avoiding 'linerslip', fitting 'shields' or 'non-return valves' into short milk tubes or liners.

4. Reduce the duration of infections by
   - — detecting clinical mastitis by examining foremilk or fitting 'mastitis detectors' into the long milk tubes
   - — giving intramammary infusions of antibiotics under veterinary supervision to clinically affected cows and keep a record
   - — Treating cows at drying off with infusions of antibiotics recommended by a veterinarian
   - — Cull cows with repeated clinical mastitis.
5. Reduce mastitis in nonlactating growing cattle or cows in the dry period
   - — Avoid using low lying grazing land and damp wooded areas where flies are common. Move cattle from pastures known to give problems with mastitis
   - — Adopt good fly control measures
   - — Treat cows at drying off with antibiotics recommended by veterinarian. All cows should be treated, alternatively treat cows that have previously shown signs of infection
   - — Note the reduction in infection is not immediate but levels fall by about 50% in one year and continue to fall in successive years.
   - — Mastitis awareness and the organisation of mastitis control

Organisation

World experience indicates that if farmers are to control mastitis successfully they require some technical and laboratory assistance. This can be provided by government agencies, cooperatives or the milk collecting dairies. The support should:

— provide an extension service (veterinary, animal husbandry and milking machine technology) to ensure that farmers receive the correct information on the best routines to follow for their environment

— report regularly to farmers (eg. monthly) giving the results of appropriate tests made on herd bulk milk to indicate the progess made in reducing mastitis

— ensure that good milking equipment, disinfectants and antibiotics are available

— investigate the causes of failure in herds adopting the recommendations but not making progress

Although laboratory support is essential it is important to recognise that mastitis is widespread and for successful control in a national herd or the herds in a cooperative it is necessary that most, preferably all, the herds carry out the routine. A control that concentrates on improving the worst 10 to 20% of herds will have little effect on the total problem of mastitis.

Tests will not often provide information that leads to special action for farmers to take that is additional to what they should be doing.

Tests for Mastitis

The tests for mastitis are either microbiological, to detect the causative pathogen, or tests for the changes in the composition of milk which occurs with the inflammation. Tests for pathogens are not required routinely but are necessary to investigate special herd problems. The tests for changes in milk composition are simpler and some may be carried out on the farm (eg. California Mastitis Test). Electronic automated laboratory tests for somatic cells in milk (eg. using Fossomatic or Coulter equipment) can be used to provide regular information to farmers.

Many other tests are also available which measure specific biochemical changes in the composition of milk (eg. Bovine serum albumin (BSA); antitrypsin test and NAGase test). The most widely used for routine screening of farm bulk milks is electronic cell counting which can readily be integrated with other tests of milk composition. When it is necessary to detect abnormal quarters on the farm a simple cowside test is useful (eg. CMT).

References

Barber F.W., "Hygienic aspects of the processing of butter, cheese, flavoured and fermented milks, and ice-cream", *Milk Hygiene*, WHO/FAO 1962.

John C.K., "Milk control on reception", *Milk Hygiene*, WHO/FAO 1962.

Newton Clark R., "Water supply on the dairy farm, (English, French, Spanish)", *Milk Hygiene*, WHO/FAO 1962.

11

Packaging of Milk

Milk is a liquid and therefore requires a container at every stage of movement from the cow to the consumer. At the early stages of dairy development the cow's udder was used as the basic container for all purposes. The cow, kept in the town stall, was brought to the customer's doorstep for milking. In some cases the milk was sold from a shop adjacent to the cowshed. In many European countries town cow-keepers could still be found after the first world war but, for reasons of hygiene and economy, they quickly disappeared. This trend seems to be unavoidable for the dairy industry worldwide and will certainly be applied to cities in developing countries where town cow-keeping still exists.

The growing demand for milk in towns and the high costs of milk production within their boundaries led to the development—probably around 1860-70—of containers suitable for various stages of marketing and distribution. These were metal cans, provided with a lid and having capacities up to about 80 litres. The introduction of this type of container (until recent years often called a 'churn') facilitated the transport by railway from rural areas to towns, thus contributing substantially to the rapid growth of milk distribution. Similar containers were also used for retail delivery to the consumer, the milk being dispensed in the street or at the doorstep into the consumer's container.

The first significant development in the packaging of milk for retail sale came at the very end of last century with the introduction of the process for sterilised milk in which the retail container, the glass bottle, formed an integral and essential part. In the third decade of this century bottling of pasteurised milk developed rapidly, first in America and soon after in Europe. The glass bottle as the retail package for milk remained unchallenged until 1933 when the first carton made of waxed paper was introduced. The development and introduction of plastic materials for packaging in the dairy industry (initially polyethylene in 1940), alone and in combination with paper, resulted in a wide range of containers, termed cartons, suitable for liquid milk.

Processing depends on the grade of milk to be manufactured following the regulations and customs of the country. Heat treatment and, in most countries, standardisation of butterfat content, are the basic parts of the processing procedures.

Heat Treatment of Milk

Heat treatment may be classified as:

- — pasteurisation
- — sterilisation (in bottle)
- — UHT (ultra-high-temperature) treatment integrated with aseptic packing.

Pasteurisation HTST (high-temperature short-time—heating at 72°C for 15 seconds) fulfils the following main objectives:

- — to safeguard public health by destroying all pathogenic bacteria
- — to extend the keeping quality of liquid milk by destroying most of the milk-souring micro-organisms
- — to ensure a product with a good keeping quality.

Sterilisation is the term applied to a heat treatment process which has a bactericidal effect greater than pasteurisation. Although it does not result in sterility, it gives the processed milk a longer shelf life. This is achieved partly by using a more severe heat treatment (about 110°C for 20-30 min) and partly by applying the treatment after the bottle is filled and sealed which eliminates the risk of contamination during packaging. As a result of the long holding time at this elevated temperature, the product has a cooked flavour and a pronounced brown colour.

UHT treatment is a process of high bactericidal effect, developed as a continuous flow process in which the milk is heated at 135°C-150°C for about two seconds only. This treatment must be integrated with aseptic packaging in sterile containers. UHT milk has less pronounced cooked flavour and no brown colour.

As a criterion for packaging requirements for pasteurised milk in general, a shelf life of several days at a temperature below 10°C can be assumed. In-bottle sterilised milk can normally be kept for weeks and UHT milk aseptically packaged can be kept for several months, both without refrigeration, provided the package is not opened. After opening, the sterility of the product is lost and the shelf life becomes close to that of pasteurised milk.

Sales of unpasteurised milk are rare in countries with a developed dairy industry and often prohibited by law. Nevertheless, there are some where purchased milk is boiled at home as a common habit even though the raw milk is of a high hygienic standard. In such cases heat treatment in the milk plant may be considered as an unnecessary expense and not required by law. This is the case in Switzerland, although even there in recent years the share of heat treated milk in the total sales is increasing.

Selection and Evaluation

The type of liquid milk produced and the consequent selection of a packaging and distribution system constitutes in many instances an extremely intricate optimising problem. The selection will have to satisfy the requirements dictated by existing economic limits, production and distribution efficiency, retailing objectives, consumer considerations and ecological aspects.

Even when the decision on the packaging system has been well founded and made with due consideration to all relevant demands, the actual application may sometimes prove to be a failure for lack of competent supervision of its application.

Returnable and single-service plastic bottles are mainly suitable for larger capacities of milk plants. Plastic bottles are normally used as single-service containers. They are usually blow-moulded from polyethylene which is often a mixture of high and low density grades. Occasionally polystyrene is used. The basic advantage of plastic bottles in comparison with glass is the lower weight of the former (about one-twentieth that of glass).

The bag-in-box system is intended primarily for milk supplies to catering establishments and has not found application to general milk distribution. Aseptic packaging of UHT treated milk in plastic sachets has not been included as this is still in the process of development and must be considered in conjunction with the UHT processing equipment employed.

The extra cost of aseptic sachet packaging equipment compared with that for pasteurised milk can be assumed to be in a similar proportion as that for aseptic and normal carton machines. It should be mentioned that a promising UHT system with aseptic filling into plastic sachets is being developed in Finland with electric heating of the milk. Commercial trials are under way.

Mlk reception, raw milk storage, standardisation, homogenisation, pasteurisation and bulk storage of milk after pasteurisation are common for all systems. Packaging operations therefore begin at the outlet of pasteurised milk tanks. It is assumed that pasteurised chilled milk is used for UHT treatment and sterilization, homogenisation (which is not always necessary) is a part of pasteurised milk processing, while storage tank types and capacities are considered as uniform though they may differ in practice. In spite of this, data resulting from this analysis can be considered as comparable within a reasonable margin of error.

Techno-economic evaluation of the systems necessitates consideration of the following essential items:

- — packaging materials and storage systems
- — machinery, equipment and area requirements in the plant
- — requirement of services
- — direct labour
- — transport to the retailer
- — retailing
- — costs and losses of packaging material and losses of milk.

These components may be considered as basic elements of what may be called divisional costs of liquid milk packaging and distribution within the definitions and limitations described above and are, therefore, a fair indication of the cost differences between the various systems.

In order to arrive at the fullest possible comparability of data, this analysis has been made for one size of package, i.e. half-litre, which is probably the most common size used in developing countries; some data are also provided for units of one litre.

Specific technical requirements for a dairy enterprise often depend on the capacity of the milk plant. For instance, the specific area for milk processing (m^2/1 000 litres) will be smaller in a 100 000 litres/day plant than in a 10 000 litres/day plant. It has been necessary, therefore, to relate the particular specific requirements to the capacity, and for this purpose five capacities have been selected, namely 10 000 litres/day, 25 000 litres/day, 50 000 litres/day, 100 000 litres/day and 250 000 litres/day. Below 10 000 litres/day a milk plant can hardly be considered as a commercial enterprise: on the other hand, in the range of capacities between 100 000 litres/day and 250 000 litres/day the specific requirements of most components become constant and continue beyond the 250 000 litres/day point. Diagrams have been used to illustrate the results, as this appeared to be the most comprehensive method of presentation.

Several assumptions concerning working time, efficiency coefficients, routeing of vehicles, etc. have had to be made. As far as feasible, they have been considered as uniform for all systems. Their values have been compared with and adjusted to real figures available from existing plants, particularly in developing countries. All factors taken into consideration in this respect are explained in the paragraphs concerned.

Specifications for basic machinery and equipment depend on the daily throughout of the division, the efficiency with which machines are utilised and on the operations in the storage and distribution rooms. In smaller units the degree of mechanisation has been limited to basic functions of the plant. In larger units operations related to storage and dispatch of the product, as well as those connected with the reception of returnable empties (crates, bottles, cans) have been considered as justifying mechanisation. The equipment listed would be usually recommended when establishing a new liquid milk packaging division. It has been assumed that, according to the practice of the majority of milk plants, the division operates two eight-hour shifts daily.

A cost evaluation of the systems, valid for developing countries all over the world, is not a feasible task. Therefore, it seemed appropriate to express first the magnitudes of the various components in physical units (m^2, kg, hours, etc) as specific values, i.e. related to 1 000 litres of milk distributed. However, a conclusive techno-economical comparison of different systems is not possible when different components cannot be added together, for instance if one component is expressed in m2/1 000 litres and the other in kgh/1 000 litres. The only common measure for all components is a currency unit, but this implies application of selected uniform rates for different expenditures (machines, construction, labour, fuel, materials, etc) and there is no uniformity of such costs all over the world. Nevertheless, an attempt has been made to assess and compare the specific divisional costs by applying uniform rates in US currency to the various cost components. They should be considered as figures indicating basically the relations between different components, and not necessarily as reflecting precisely the real costs of any given project. In order to arrive at estimates of actual costs the methods of calculation given can be applied, but actual rates should be used.

There are numerous different designs of machines and kinds of packaging material for liquid milk packaging on the market. With very few exceptions, the equipment and packaging material for each of the systems are produced by a number of companies. They differ in design, capacities, weight, dimensions, etc., but many of them fulfil equally well the hygienic, technical and economic requirements for use in modern dairying.

Characteristics of Packaging Systems

Returnable Eontainers

The basic features of systems using returnable containers are the collection of empties and washing prior to re-filling. Differences in operation times and capacities of the various machines involved make intermediate storage necessary. Storage of unwashed empties is normally essential and may extend overnight so that washing and filling operations can begin next morning before the day's supply of unwashed empties arrives. Storage of washed cans is permissible as they have lids but storage of washed bottles is extremely bad practice because they are unsealed and therefore liable to contamination. Normally storage must be provided for filled cans and bottles to give flexibility in the distribution arrangements. For pasteurised milk this must be refrigerated.

Bottle washing, filling and capping machines should be of matching capacity, otherwise the labour-intensive operations of decrating and crating, as well as unstacking and stacking, would have to be repeated unnecessarily. This problem does not arise with cans, since they are not crated and may be easily stored empty after cleaning.

The required storage area, both for empties and for product, depends on the operation schedule of the plant which in turn is affected by the relation between required dispatch capacity and the capacity of filling machines in operation.

Pasteurised milk in bottles

Bottles with wide necks (36 to 40 mm), suitable for sealing with aluminium foil caps made in situ from reeled strip, form the most common system for packaging of pasteurised milk in returnable containers. The bottles are placed into crates, formerly made of galvanised steel wires or strips and nowadays usually of plastic. The crates have internal divisions so that the bottles are not in contact with one another to minimise risk of breakage. They are designed to interlock, so that a stable stack can be built. For manual handling crates with filled bottles are stacked five and six high for one-litre and half-litre bottles respectively; with empty bottles two to three crates more are put in one stack. Crates usually hold twenty half-litre and twelve to fifteen one-litre bottles.

The stacks can be palletised (Standard Europallets 800 × 1,200 mm are applicable) and moved by fork lift trucks or manually or by hand trolleys without palletising. There are no international standards for the dimensions of pasteurised milk bottles and crates.

The floor area occupied by one stack of crates is thus about 0.15 m^2, equivalent to a milk storage capacity of 400-470 1/ m^2 depending on bottle capacity and stack height. For crated empty bottles the equivalent storage capacity is about 30% greater. The initial high cost of a glass bottle prevents single-service use for pasteurised milk. The effective cost depends on the number of times the bottle can be re-used (trippage) which in turn is determined primarily by the effectiveness of the bottle recovery system and the ability of the bottle to withstand breakage. In the UK the weight of a 1-pint bottle (568 ml) is about 12 oz (340 g) and costs about 4.25 p (1977). In recent years a new design has been introduced weighing about 8 oz (227 g) costing about 15% less but giving the same, or only slightly less, trippage.

Sterilised milk in bottles

Bottles used for in-bottle milk sterilisation have narrow necks (26 mm) so that a more effective seal can be made. Prefabricated crown seals are used to seal the bottles. These bottles must be able to withstand not only mechanical shocks during handling, but also thermal shocks during sterilisation and, even more, during cooling. As the milk in the bottle is heated and expands during heating more than the bottle, the air above the milk becomes compressed and the pressure inside the bottle exceeds the external pressure. The contraction of the milk as it cools below the filling temperature results in a vacuum in the space above the milk. This vacuum may encourage contamination through the seal between the bottle and the cap. It is therefore important that the seals be fully airtight.

As can be seen from the above, the demands for a bottle for this process are particularly high, and it must therefore be heavier and more expensive. The bottles are packed in crates as in the method described for pasteurised milk bottles. The stacking system and the resulting storage area required are also similar. As with pasteurised milk, dimensions differ from country to country but a neck diameter of 26 mm is universal.

Pasteurised milk in cans

The aluminium milk can has proved very satisfactory in service and, since the beginning of the second half of this century, has rapidly replaced the previously-used tinned mild steel can. In recent years high density polyethylene cans began to be introduced in a number of countries, but have not proved popular for various reasons.

The most common are cans with lids which do not require rubber gaskets, an adequate seal being achieved with sunken grip or mushroom lids. Because of mechanical washing problems lids attached to the cans by chains are no longer used. Through simple arrangements at the lid ring lead or other seals can be applied to make the contents of the can pilfer-proof. The cans may be palletised, but more often floor conveyors are used. Full cans are stored in one layer, thus allowing about 320 to 360 litres of milk to be stored per square metre excluding access space. Empty cans, after washing, are stacked in layers horizontally, up to the height of 1.5. m. For storing and moving washed empty cans simple trolleys on which the cans can be stored in 4-5 layers are very useful; for instance, about 20 cans each of 40 litres capacity can be stored on trolleys about 1 700 × 700 mm with a supporting frame made of a 1/2" pipe.

Single-service Containers

The common feature of single-service containers is that after emptying they are discarded. This fact has a significant impact on the milk plant construction, organisation, and on the economics of the whole enterprise. There is no collection and washing of the milk packages—only crates are collected and washed, but even these may be replaced by single-service delivery wraps, trays or boxes. Palletisation may be applied as in the case of returnable containers. Intermediate storage of packing material and filled packages is required and this must be provided in the plant.

Two basic types of single-service containers are considered in this analysis, i.e. cartons and plastic sachets. Cartons are usually made in one of the shapes illustrated in Fig. 1.

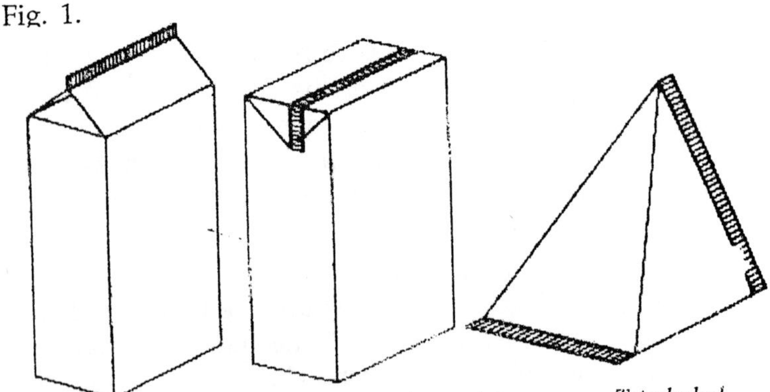

Rectangular containger wth a gable top — *Rectangular containger with a flat top* — *Tetrahedral package*

Figure 1. Basic shapes for milk cartons

The material used in each case is polyethylene-coated paper-board; in aseptic versions an aluminium foil lamina is normally incorporated. Pre-formed cartons or pre-cut, single-piece blanks pre-creased ready for forming into a container must be used for the container with a gable top and these generally are more expensive. The other method of making a carton is to form a tube from a reel of material, seam it longitudinally, fill with milk and then make transverse seals. Alternatively the tube may be cut into lengths which are formed into cartons before filling and sealing.

Plastic sachets are usually pillow-shaped and made of low density polyethylene film. They may be reeled single or double film or lay-flat tube, the latter avoiding the necessity of making the longitudinal seam in the packaging machine. The material should be coloured to reduce light transmission.

The sequence of operations when packaging into single-service containers comprises forming the container, filling and sealing, storage of the packaged product and dispatch to wholesale and retail outlets. As with packaging in glass bottles the storage space requirements must be related to the process schedule but there are important differences. Otherwise there is no constraint on the beginning of the day's packaging operations.

The distribution of food is changing rapidly in the industrialised part of the world. Traffic routes from processor to retailer are becoming increasingly crowded, sales through supermarkets are demanding changes in the systems of packaging and developments aimed at making this distribution feasible and manageable in large concentrations of consumers are taking place constantly. There is no doubt that single-service containers respond better than returnable ones to the demands of this modern trend.

Milk distributed in single-service containers in returnable crates relieves the retailer from the necessity of collecting empties from the customer but the need for returning crates to the processor still exists. This may not create too many problems since empty crates may be collected by the processor at the time of milk delivery, provided two sets of crates are put in operation. A completely one-way packaging system requires that the single-service containers are wrapped in non-returnable material, thus forming transportable units.

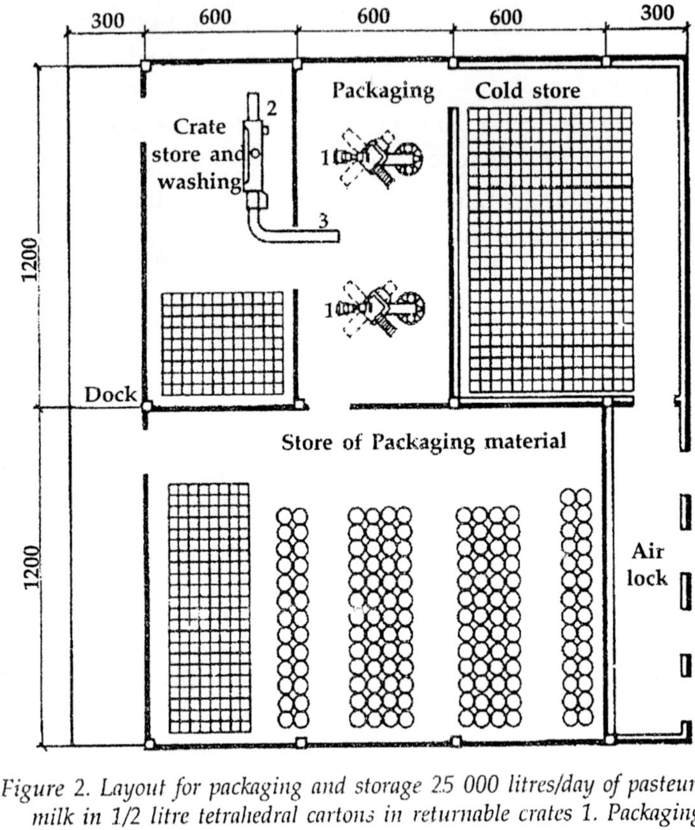

Figure 2. Layout for packaging and storage 25 000 litres/day of pasteurised milk in 1/2 litre tetrahedral cartons in returnable crates 1. Packaging machines; 2. Crate washer; 3. Crate conveyor

Pasteurised milk in cartons. Tetrahedral cartons made from polyethylene laminated paper board and packed in hexagonal plastic crates have been chosen as the model for analysis of a system of pasteurised milk packaging. The cartons are produced continuously from a roll of plastic-coated paper which is shaped and sealed into a tube. The tube is filled continuously with pasteurised milk.

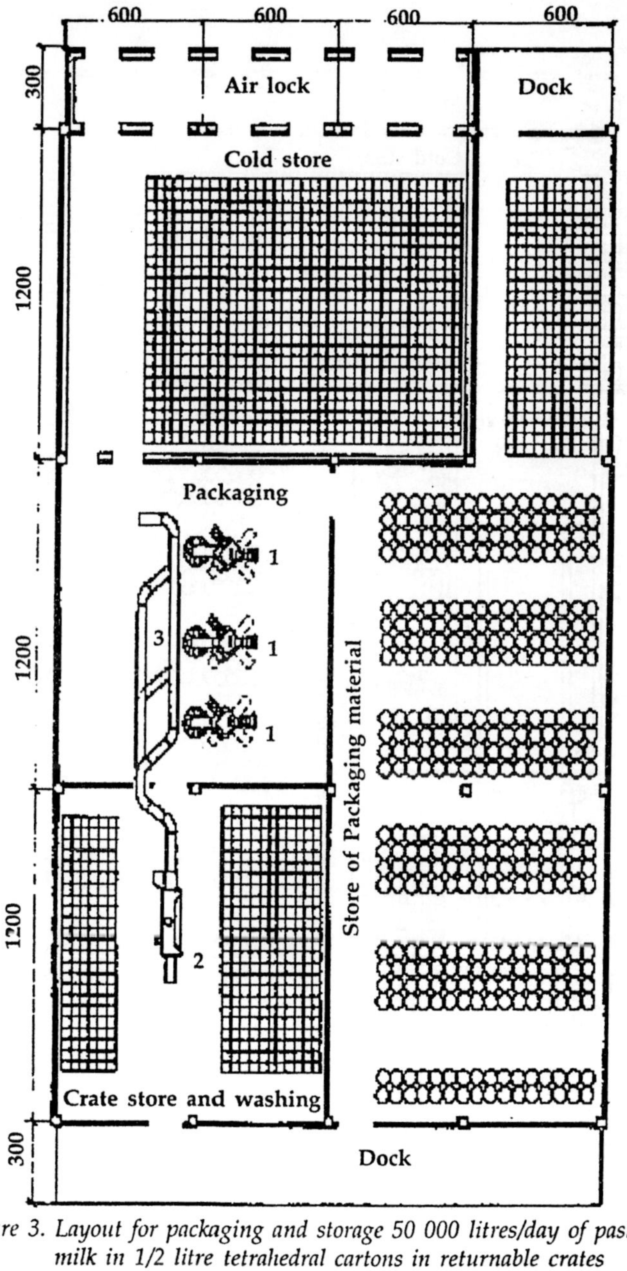

Figure 3. Layout for packaging and storage 50 000 litres/day of pasteurized milk in 1/2 litre tetrahedral cartons in returnable crates
1. Packaging machines; 2. Crates washer; 3. Crate conveyor

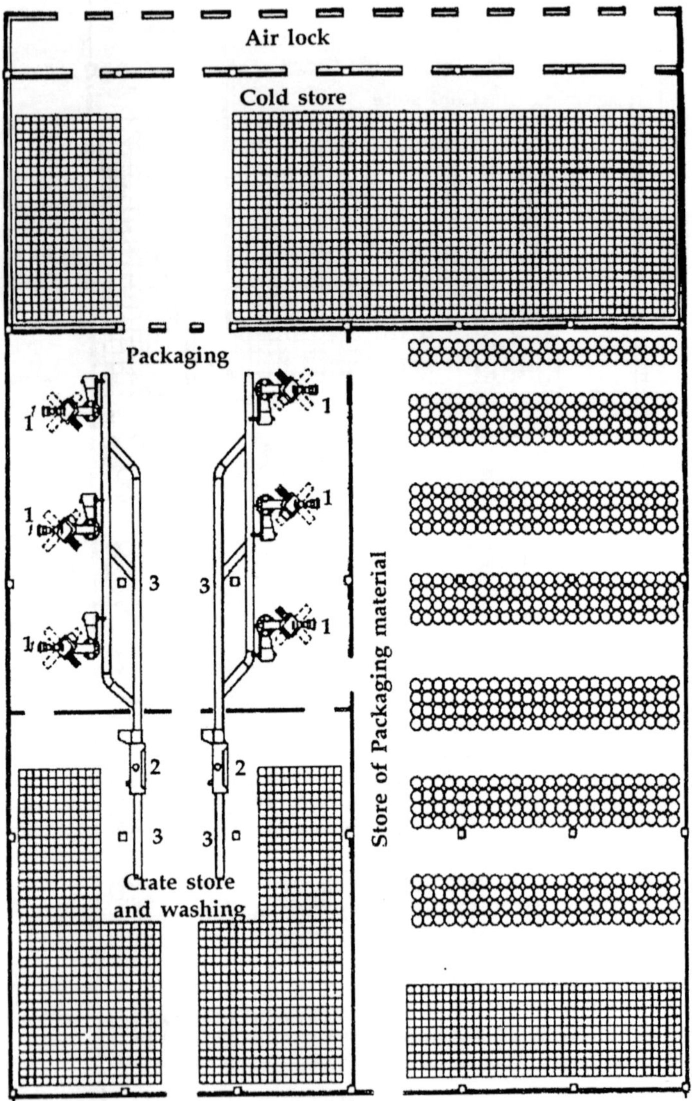

Figure 4. Layout for packaging and storage 100 000 litre/day of pasteurized milk in 1/2 litre tetrahedral cartons in returnable crates
1. Packaging machines; 2. Crates washer; 3. Crate conveyor

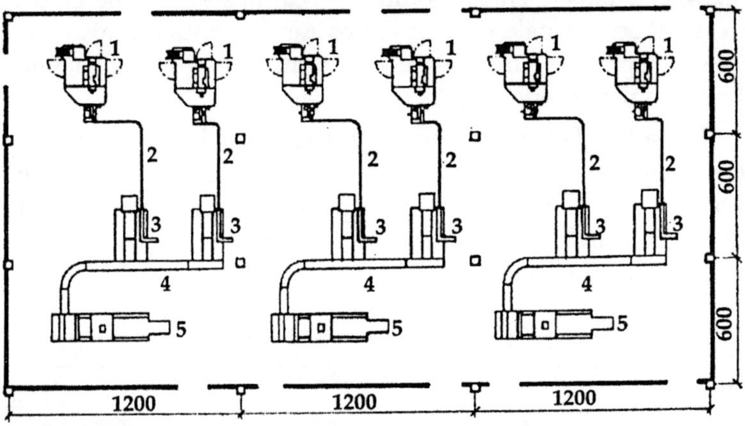

Figure 5. Example of layout for machinery room for aseptic packaging 100 000 litres/day of UHT milk in 1/2 litre single-service rectangular cartons in non-returnable wraps 1. Packaging machines; 2. Package conveyor; 3. Collating and packing on trays; 4. Conveyor; 5. Shrink wrapping

Transverse seals are made alternately at right angles below milk level so that there is no headspace and the shape of the package is a tetrahedron.

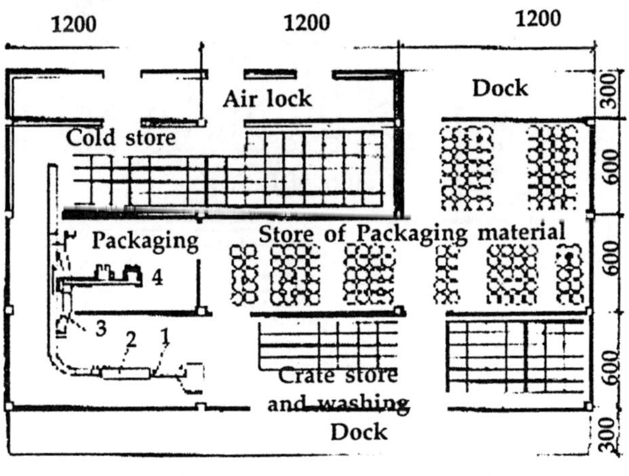

Figure 6. Layout for packaging and storage 50 000 litres/day of pasteurised milk in 1/2 litre plastic sachets in returnable crates 1. Crate conveyor; 2. Crate washer; 3. Crate filling; 4. Packaging machine

The packages are separated by guillotine and placed in hexagonal plastic crates holding 18 cartons each. The crates are stacked on pallets. The crates (and eventually also the pallets) have to be transported from the retail centres back to the milk processing plant.

For a 250 000 1/day plant packaging in 1/2-litre cartons it represents only 2 days production. Depending on the location of the milk plant and the paper supplier and the most economic purchasing quantities it may be necessary to carry two to three months stock.

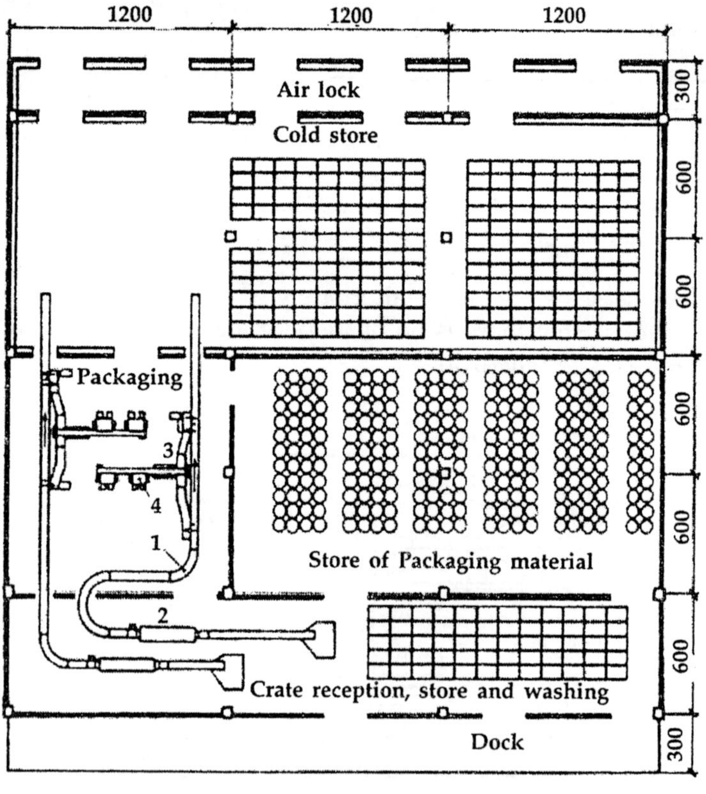

Figure 7. Layout for packaging and storage 100 000 litres/day of pasteurised milk in 1/2 litre plastic sachets in returnable crates 1. Crate conveyor; 2. Crate washer; 3. Crate filling; 4. Packaging machine

Figs 2 to 8 give storage areas for packaging material equivalent to about 60 days stock for the 250 000 litre plants and 70 to 100 days for the smaller capacities.

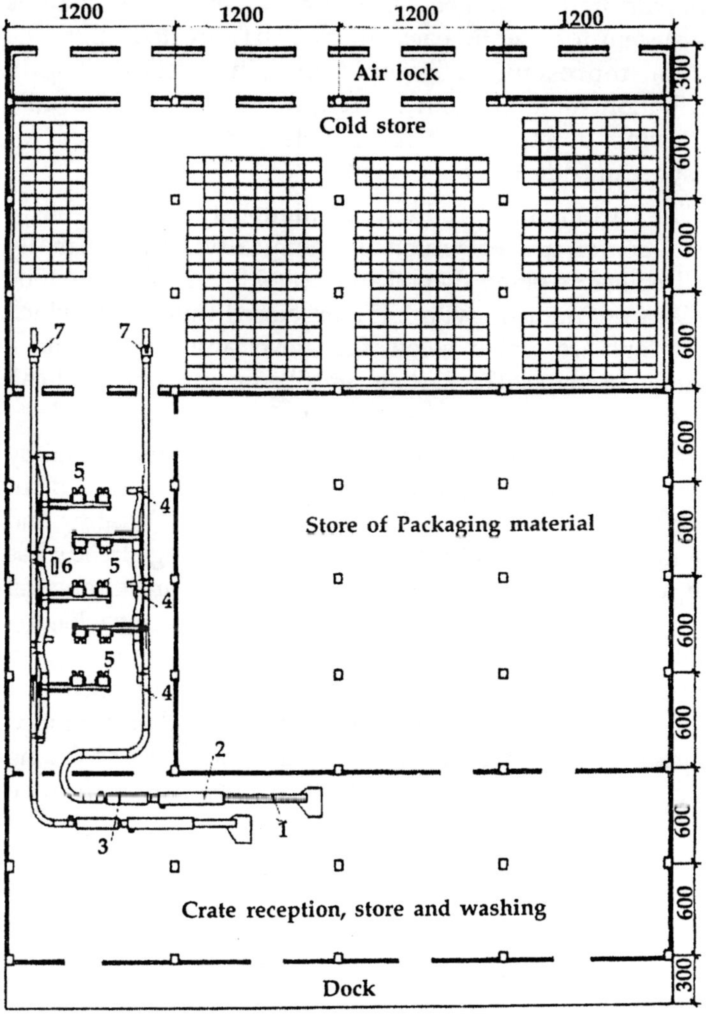

Figure 8. Layout for packaging and storage 250,000 litres/day of pasteurised milk in 1/2 litre plastic sachets in returnable crates 1. Crate conveyor; 2. Unstacker; 3. Crate washer; 4. Crate filling; 5. Packaging machines; 6. Control panel; 7. Restacker

UHT milk in cartons

Rectangular cartons made from polyethylene laminated paper board in shrink-on wraps strengthened by corrugated cardboard trays have been chosen as the model for analysis of a system for aseptic packaging of UHT treated milk. The system represents a completely single-service milk distribution method, with no return transport to the milk processing plant, except for pallets if they are used for transport outside the milk plant.

The cartons are produced continuously from a roll of plastic-coated paper which is chemically and thermally sterilised before being shaped and sealed into a tube. The tube is filled continuously with UHT processed milk, after which the cartons are sealed below fluid level and formed into a rectangular shape. The cartons are filled completely and can be stacked.

Out of several distribution alternatives for the method by which the single-service containers are collated in transportable units, the most often applied are: (a) crating and (b) wrapping in non-returnable materials. Crating already has been analysed in relation to pasteurised milk in tetrahedral cartons. For one-way (non-returnable) transport packaging shrink-film is most often used. This wrapping can be applied manually or mechanically. When machines are used, the cartons are fed to the marshalling station and then wrapped in shrink-on film. After having passed through the shrinking tunnel the packages are loaded on to pallets and are then ready for distribution.

This system is used where short storage times and short distribution distances are involved. However, the film does not give much mechanical support to the packages and cartons wrapped in shrink-on film are often subject to damage, leakage and contamination, particularly when handled manually without the use of pallets and other auxiliary equipment. Special (perforated) shrink-films may be required in areas with high humidity to prevent condensation. A much safer, although more expensive, system is that by which a group of twelve rectangular cartons filled with milk is placed first on a flat tray blank. The sides of the blank are then folded up tightly around the cartons so utilising to the full the pressure-absorbing ability of the filled cartons.

The packed corrugated tray is then wrapped in shrink-on film which not only affords protection against dust and moisture but also presses the sides of the tray in towards the cartons. This brings about a substantial improvement in the pressure-absorbing ability of the corrugated tray, at the same time giving a compact transport packing capable of resisting the strains and stresses encountered in the course of distribution. The units can easily be handled manually.

This system of packaging the cartons in one-way transport units by wrapping them in shrink-on film after having strengthened the unit of 12 cartons by a tray of corrugated cardboard is used for UHT milk basically meant for longer storage and longer distances of transport (such as once-a-week delivery to a consumption centre from a distant milk plant).

It seemed to be appropriate to take into consideration and present data for this method of wrapping the cartons into one-way strengthened transportable units so as to arrive at indications for the costs—by today's standards—of the most modern liquid milk processing and distribution system. Up to about 600-700 litres of milk per square metre can be stored when pallets are used (without access).

Pasteurised milk in sachets. Pillow-shaped sachets with a longitudinal seam made from reeled low-density polyethylene film have been chosen as the model for analysis of the system. The film is shaped and welded into a tube. The tube is filled with pasteurised milk from a small balance tank, where the level is kept constant by means of a float. A timer-controlled pneumatically-operated valve is used to dispense constant quantities of milk. The transverse seals are generally made above milk level. The packages are separated by guillotine and placed in rectangular plastic crates holding 20 packages each and the crates are then palletised. The same width and thickness (90 mm) is used for both 1-litre and 1/

2-litre packages; capacity is varied by varying the distance between transverse seals. It is important that the film is free from pinholes or micropores.

PACKAGING EQUIPMENT AND PLANT AREAS

The key item of equipment is the packaging machinery, the capacity of which is derived directly from the required daily throughput. The capacity of all other items of equipment (the processing equipment being appropriate to the type of milk to be produced) must be related to this. A more detailed specification of these items is not necessary for the purpose of this analysis.

Different makes of equipment for the same purpose inevitably show differences in capacity, dimensions, cost and other details. It may well be that one make is more suitable than another at one particular plant throughput while a different make has the advantage at some other throughput. It has been necessary, therefore, to make an arbitrary selection to achieve consistency in data comparisons.

In selecting the basic machines it is necessary to consider whether the required capacity should be covered by one machine or by two of a smaller size. One machine of the full capacity required is cheaper in all respects but two of half capacity each is a much safer solution, particularly where spare parts replacement and general maintenance may create problems. Wherever feasible two half-size machines should be preferred to one full-size.

The indicative value of the cost component related to machinery and equipment has been calculated separately for machines owned by the plant and those obtained on lease because some makes of packaging equipment are available only on the latter basis.

The cost component, Z, has been computed for both owned and rented machines by the following equation

$$z = \frac{C}{365a}(D + B + E) \text{ in US\$ / 1000 litres}$$

where C = value of owned (C_1) or annual expenses on rental (C_2) for rented machines in US$

a = throughput in thousand litres of liquid milk per day

D = 'depreciation' (12.5% of C, for owned machines and 100% of C2 for those on lease)

G = rate of annual maintenance costs (10%)

E = rate of annual expenses on insurance and taxes (2.5%)

The resulting magnitudes are:

— for owned machines Z_1 = about $0.001 \dfrac{C_1}{a}$

— for machines on lease $Z_2 = \dfrac{C_2}{a}$

It follows that $Z = Z_1 + Z_2$

Having estimated the areas required for machinery rooms and stores the layouts have been designed to conform with the internationally agreed structure modules of 12 m span and 6 m centres for structural beams. The application of these module dimensions gives a better comparability of different sizes and types of buildings but this may leave more space than is required for a particular purpose, particularly in plants with smaller throughputs. Although milk packaging and storage do not represent all the milk plant operations the layouts have been prepared as regular rectangles in order to make the sequence of movements more clear to the reader.

Transport

The type of liquid milk distributed, the packaging system, the number and location of retail centres, the distance between the retail centres and the plant or depots, and the type of vehicles are major factors affecting the techno-economics of transport. In some areas milk is sold only in the morning, in some twice a day and in some others—particularly where refrigeration is available or UHT or sterilized milk are popular—throughout the day. These purchasing habits of the consumers have also to be considered when assessing transport requirements.

In order to work out a comparative analysis of the techno-economics of transport of different types and quantities of liquid milk, simplified distribution models had to be designed. The analysis is made under these model conditions.

The topographic model of the area of distribution is shown in Fig.9. It is assumed that a daily production of 10,000 to 100 000 litres will be distributed within a square 10 km × 10 km and for 250 000 1/day the area to be served will be a square 20 km × 20 km. Focal points around which 1 000 1/day are sold are marked with a cross. They do not necessarily indicate the locations of retail shops, but rather represent the topographical designation of centres of milk sales areas.

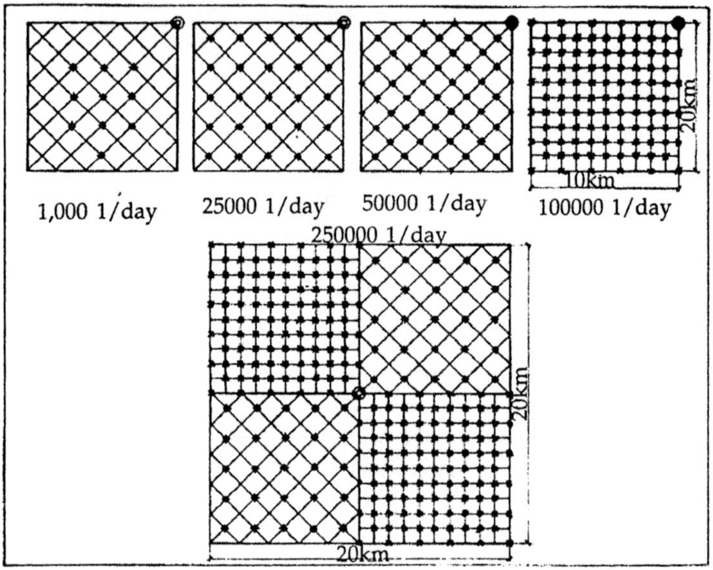

Figure 9. Models of distribution areas (milk plant; + focal points of sales areas)

This implies the assumption that the additional distance which the milk delivery vans have to cover around the focal point in order to reach the actual retail shops is negligible as compared to the distance between focal points and the plant. The analysis of delivery vehicles routeing has been done separately for each capacity and each milk packaging system. An example of this analysis is shown in Fig. 10.

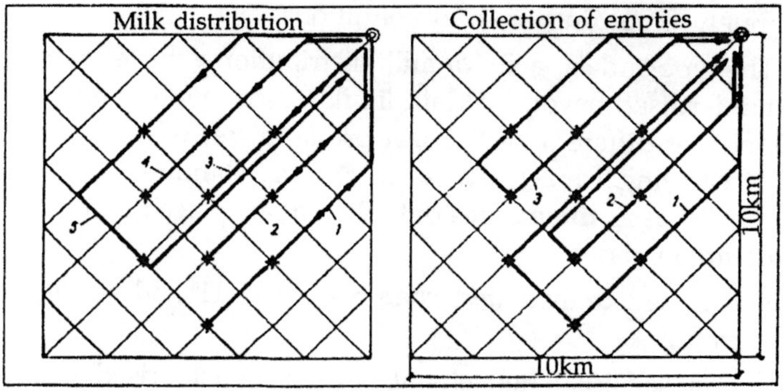

Figure 10. Model distribution routes for 10 000 litres/day of pasteurised milk in 1/2 litre glass bottles

The routeing indicated for the transport is not necessarily the shortest. In a real area access to roads, density of traffic, local regulations, etc. will affect the selection of routes and will certainly deviate from the theoretically shortest. In the model presented, routeings have been selected following a common pattern thus permitting comparisons between different systems and capacities. It was found that the quantity of milk distributed has little influence on specific distance expressed in km/1 000 litres. For practical calculations the specific distance may be considered as constants for each of the distribution systems.

Transport costs of UHT milk are affected by the fact that the expense of UHT treatment and aseptic packaging can only be justified for markets which require a shelf-life of the product of at least 2 months. Such markets include those where distribution involves very long distances or times, i.e. where the processing plant is at a long distance from the retailer. Transportation costs from the plant to a depot in the consumption centre could not be included in the calculations presented since no relative model conditions could be drafted. UHT milk is to be considered as covering transport operations from a depot, instead of from a milk plant. Additional transport costs from plant to depot should be calculated separately according to local conditions.

The type and capacity of milk distribution vehicles cannot be standardised beyond certain limits, since requirements are different for different systems. A prime-mover with a chassis of 5 t carrying capacity was selected as standard. On this standard chassis three types of body may be placed according to requirements:

i. a closed uninsulated chamber for UHT and sterilised milk;
ii. a closed insulated chamber for all pasteurised milks, except for delivery to vending machines;
iii. an insulated milk tank for deliveries of pasteurised milk to vending machines.

The milk-carrying capacity of the chassis was calculated excluding the weight of the appropriate body. It was also assumed that the required platform area may be chosen without affecting the costs of the chassis.

Standard timings of milk delivery to retail centres have been adopted as follows:

i. Pasteurised milk to vending machines, UHT milk and sterilised milk are delivered during two 8-hour shifts, since the products can be kept safely in the retail centre.
ii. Pasteurised milk is delivered to retail centres within one 8-hour shift. Empty bottles and cans are collected during the second 8-hour shift.
iii. Empty sterilised milk bottles and crates and the crates for pasteurised milk in single-service containers are collected during delivery trips.

Simultaneous distribution of milk and collection of empties for pasteurised milk in bottles and cans is generally considered as impractical in conditions prevailing in developing countries as the lack of refrigeration necessitates sales during a short time and whole-day storage of uncollected empties at the retail centre is undesirable.

Costs and Losses of Packaging Material and Milk

Packaging material sold together with the product and losses of packaging material and product during packaging, handling in stores and transportation, represent a substantial part of packaging costs. They may be classified as follows:

A. Packaging material

1. Value of returnable bottles lost or broken during distribution.
2. Value of returnable bottles damaged during washing, filling and handling at the milk plant.
3. Loss, depreciation and maintenance of cans.
4. Value of single-service containers purchased by the customer together with milk.
5. Value of single-service containers damaged during packaging and transportation.
6. Loss, depreciation and maintenance of crates.
7. Value of packaging material for single-service wraps.

B. Milk

-- Losses due to over-filling or measuring and wastage through spillage and breakage during packaging and transportation.

Packaging Material

Pasteurised milk in glass bottles. The impact of the cost of the bottle on the retail price depends on the trippage of the bottle. With a reasonable quality of bottles it seems to be fair to assume that a glass bottle for pasteurised milk can be re-used at least 20 times which means that about 5% of the value of the bottle covers items as specified above.

The cost of a bottle is assumed to be US$ 0.08 and to be US$ 5.00 for a crate. The approximate costs and losses of packaging material for 1 000 litres of milk may therefore be estimated as:

Sterilised milk in glass bottles. The trippage of bottles for sterilised milk is lower and re-utilisation is assumed to be 15 times. The cost of a bottle is taken as US$ 0.10. The cost and utilisation of crates could be assumed to be the same as for pasteurised milk. The approximate cost for 1 000 litres of milk is therefore:

Pasteurised milk in 40-litre cans. The cost of a can, including maintenance, can be estimated at US$ 40.00 and it is assumed that it can be re-used 1 000 times. The approximate cost of the packaging material for 1 000 litres is therefore:

$$1000 \times \frac{40.00}{40 \times 1000} = \text{US\$ } 1.00$$

Pasteurised milk in tetrahedral cartons in returnable crates. The cost of 1 000 cartons (equivalent to 500 litres of milk) is about US$ 15.00 i.e. US$ 30.00 per 1 000 litres. Loss during packaging and transportation (about 1%) increases the cost to US$ 30.30.

UHT milk in rectangular cartons in single-service wraps. The cost of 1 000 cartons (equivalent to 500 litres of milk) is about US$ 30.00 i.e. US$ 60.00 per 1 000 litres. Loss during packaging and transportation is about 2% which increases the cost to US$ 61.20 for 1 000 litres. The material for single-service wrapping is about US$ 100.00 per 1 000 packs, holding 6 000 litres.

Pasteurised milk in plastic sachets in returnable crates. The cost of polyethylene film in three colours can be estimated at US$ 2.00 per kg which is equivalent to about 250 1/2-litre sachets or 125 litres of milk. Packaging and transportation damage is rather high and should be estimated at 4%. The resulting approximate cost of sachets is therefore about US$ 16.34 per 1 000 litres. The price of a crate for 10 litres, usable for 1 000 trips, is about US$ 3.50.

Losses of Milk

The liquid milk processing and distribution systems may be

divided in four groups with respect to filling/measuring accuracy:

— filling of glass bottles
— filling of single-service containers
— filling of milk cans
— dispatch in tankers to automatic vending machines

The accuracy of filling returnable bottles and single-service containers directly affects the consumer whereas filling of cans or tankers does not. In the latter case the consumer receives the milk measured out of the can or dispensed by the vending machine and it is therefore the accuracy of the means of dispensing at this second stage which affects the consumer.

Glass bottles are filled by the machine to a level which is at a constant distance from the top of the bottle and, therefore, the accuracy of filling depends mainly on the appropriate standards and uniformity of the bottles. The tolerance for bottles may permit differences in the filled quantity as high as -15 ml + 7.5 ml, which in a 1/2-litre bottle may mean -3.0% + 1.5%. Usually control by check tests on new bottles delivered and well-managed filling operations keep the actual tolerance in filling to within ± 1%, without significant losses to the milk plant due to overfilling.

Single-service containers are normally filled with a very high accuracy, normally within the range of ± 0.2 to 0.4%, also with practically no losses to the plant due to over-filling. Milk cans and road tankers are most often filled by means of milk flowmeters which operate within ± 0.3% if properly maintained. Measuring from the can depends on the dispensing system used. Volumetric dispensing by a metering pot is very inaccurate and the customer may receive, particularly when the can is less than half full, less than 90% of the quantity paid for if measuring is not done very carefully. A manually-operated mechanised dispenser mounted on top of the can is a better alternative. This operates with an accuracy within ± 1.0%. Automatic vending machines, if properly maintained, dispense the milk within a tolerance of ± 1.0%, normally without losses due to over-dispensing.

The wastage of milk during packaging and transportation may differ considerably depending on the quality of the packaging material, maintenance of machines and management of the packaging division, storage and transport. Not all the milk from damaged packages is wasted: a substantial proportion, basically from packages damaged during the packaging process, is usually recovered. Even in well-managed plants with supplies of packaging material of appropriate quality the losses are seldom less than 0.5%. They may become significantly higher particularly with sterilised milk in bottles when the bottles do not withstand the thermal and mechanical shocks during sterilisation and after-cooling. They may also become very high with pasteurised milk in plastic sachets because of faulty packaging material and the general fragility of the package.

Keeping quality

Pasteurised milk, however packaged, has a very limited keeping quality because spore-forming micro-organisms survive the heat treatment and germinate later to cause the milk to sour. Even if it is stored at the milk plant, transported and stored at the retail centre or shop at a temperature below 4°C it should be sold to the consumer within about 24 hours of processing. Milk dispensed from cans or automatic vending machines is unlikely to be kept cool after sale and its keeping quality will be greatly influenced by the hygienic condition of the consumer's container. The milk cannot be expected to remain sweet for more than a few hours in warm climates unless it is boiled forthwith, as is a common practice under such conditions.

If the purchaser of pasteurised milk packaged at the milk plant and handled under refrigerated conditions can place the package unopened in a domestic refrigerator within a short time it would be reasonable to expect the milk to keep several days before opening. Once the package is opened the onset of souring of any milk not used immediately will still be delayed for some hours if it is re-placed in the refrigerator, especially if it is retained in the original package. When the consumer has no access to a refrigerator it is usually necessary to use the milk on the day of purchase.

In-bottle sterilised milk at the completion of processing contains very few residual micro-organisms, indeed some bottles may be sterile. Refrigeration is therefore unnecessary and it is generally safe to delay sale of the product for up to 7 days. This may have advantages for the milk plant and the retailer though it must be remembered that the milk represents locked-up working capital until it is sold. However, these two factors make it possible to distribute the milk to far greater distances, though this is limited in practice by the need to recover the bottle. This does not apply if single-service plastic bottles are used. After sale, in-bottle sterilised milk should have a shelf life at ambient temperature of a further 7 days before opening the package. After opening, unused milk remaining in the bottle may still remain sweet for several days provided the bottle is hygienically re-closed after each use.

The UHT process combined with aseptic packaging is such that the packaged milk is virtually sterile. A practicable hygienic standard is that there should be not more than one organism in 1 000 packages. Thus if no micro-organisms are present there can be no bacterial spoilage and the keeping quality on this account is unlimited. In practice chemical instability determines the shelf life and this can be 4 to 6 months before consumption. After opening the package the same considerations as for in-bottle sterilised milk apply. Typical markets for UHT milk are thus those which need, and can pay for, this long life. They include supply to very distant markets, including export, for ships and aircraft and, in some cases to local consumers who wish (and can afford) to keep an emergency stock in ambient temperature storage.

Returnable versus single-service containers

It will be obvious that the success of a system involving re-use of containers—normally glass bottles—depends entirely on the efficiency of bottle recovery. This is achieved in one of three ways. Firstly, the milk may be delivered to the customer's doorstep and empty bottles recovered at the same time. This system has the advantage that it can work in the customer's absence, payment being collected, say, once per week. Secondly, travelling shops may tour the area with frequent stops in each street making their presence known by an audible signal.

The customers come to the vehicle bringing empty bottles on which they have already paid a deposit equal to the cost of the bottle. If they require more full bottles than the number of empties they return, extra deposits must be paid. If they require less a corresponding credit is allowed. Thirdly, milk may be sold only from shops to which the customer must go to purchase milk. A similar arrangement whereby the consumer pays a deposit on the bottle is usually operated.

Where one of these methods or a combination of them can be operated it is almost certain that for pasteurised and in-bottle sterilised milk the returnable bottle is the cheapest form of retail packaging. At the present time, countries using glass bottles for more than 50% of packaged pasteurised milk include Bulgaria, India, Japan, Malta, New Zealand, Poland, South Africa and the United Kingdom.

Nevertheless, single-service packages have certain advantages as evidenced by the fact that they are widely, if not exclusively used for pasteurised milk in Austria, Brazil, Canada, Switzerland, Sweden, Denmark, Germany, Finland, France, Israel, Norway and USA. For UHT milk, single-service containers must be used as there is no commercially available system for aseptically filling glass bottles. From an organisational point of view, neither the milk plant nor the retailer has to worry about container recovery and the capital cost of equipment and space for making the container fit for re-use are eliminated.

The single-service container is much lighter in weight and generally occupies less space, both of which advantages are attractive to the consumer particularly if he has to purchase from a shop. As has been shown, all the single-service packages are more expensive than the returnable bottle where this is practicable. The most competitive is the sachet because it uses the least weight of material and the packaging operation is simple but this is offset to some extent by the unattrative nature of the filled package and its vulnerability to damage.

From the point of view of availability of materials most countries can produce glass from their own resources whereas in developing countries paper/plastic laminates and the high technology required to produce them must generally be imported with a corresponding need for foreign currency. In some cases disposal of used packages may also present a problem.

These characteristics of the various methods of processing and packaging liquid milk must be borne in mind in relation to the character of the market it is intended to serve and the habits of the milk-consuming population. It should be noted also that the taste and colour of pasteurised milk is different from either in-bottle sterilised or UHT milk and all are different from raw or boiled milk. Where milk is being introduced virtually as a new food these differences do not matter but where the consumer is already accustomed to a particular type time may be needed for a change to be accepted.

Retail Packaging vs Sales from Bulk

Several factors have to be considered when comparing the merits of packaged milk with those of milk sold from bulk. Firstly, as has been shown above, only pasteurised milk can be distributed either packaged or from bulk. Both of the other types (sterilised and UHT) cannot be sold without packaging. Secondly, for reasons also discussed above, the keeping quality of pasteurised milk depends on the temperature at which it is kept after processing. If it leaves the milk plant cold store at a temperature well below 10°C, is transported to the retail shop in insulated vans and is kept by the retailer under refrigerated conditions, the keeping quality of the packaged product after sale is almost always likely to be better than if the milk is dispensed into the customer's container. In addition, packaged milk is generally measured accurately and is safe from adulteration, which is not the case when sales are effected from cans.

If at any stage of operations the temperature of the milk is allowed to rise its keeping quality decreases at a rate depending on this rise and on the time for which it is exposed to ambient temperatures before sale. The system of distributing milk from bulk through automatic vending machines offers a continuity of the 'cold chain'. The temperature of the milk dispensed into the customer's container is low which gives it an advantage over pasteurised milk, packaged or in cans, sold under unrefrigerated conditions.

One of the major drawbacks of sales through vending machines is the necessity to rebuild or construct centres designed to accommodate the machines. This may often be difficult in crowded cities. Besides, both the known designs of the machine (as applied in Mexico and India) are available only in capacities of 2,000 and 1 000 litres. Simplification and further development of these machines is desirable in order to make their application more flexible and suitable for varying conditions and requirements. The experience gained by their application, particularly in India, seems to indicate that the system enjoys full support from the consumers, both in the lower and higher income strata of the society.

References

Bailey, J.C. & Brace, A.W., "Experience with aluminium milk cans", *Journal of the Society of Dairy Technology*, 1956.

Ellis, J.G., "Aseptic processing and packaging of milk", *Journal of the Society of Dairy Technology*, 1977.

Hall, H.S., "Retail packaging of liquid milk", *Journal of the Society of Dairy Technology*, 1973.

Hall, H.S., *Standardized Pilot Milk Plants*, FAO/ICP Rome, 1976.

Poulsen, P, Ronkilde, "Selecting packaging systems", *International Dairy Federation Annual Bulletin Document*, 1976.

12

Milk Production and Marketing: Challenges and Opportunities

Asia with 60% population of the world accounts for only 20% of the world's milk production. The total world milk production of about 500 million tonnes for a population of about 5 billion people amount to an average annual per capita milk consumption of 100 kg. Asian milk production is only about 27 kg per capita and the consumption is about 30 kg per capita. Some 10% of milk consumed in Asia is imported, with local milk production accounting for 90% of the milk consumption.

Other sources of animal protein like meat and eggs are also not available to the Asian populations in the desired quantities. While the world average annual meat and eggs consumption is about 40 kg per capita, Asians consume only about half of the world average i.e. 20 kg per capita.

The rate of growth of milk consumption in Asia has however been around 5% over the past decade as compared to the world average growth rate of about 1%. The consumption of meat and eggs has also doubled over the past decade. The Asian livestock production index has been around 134 (1980:100) as compared to the world index of 104 over the 1980 – 1990 decade.

The redeeming factor, however, is the availability and higher consumption of beans and pulses which supply the bulk of the protein requirements.

Whilst countries like Pakistan and Japan have a per capita availability of milk which is either close to or exceeds the world average, the world's most populous country, China consumes only about 6% of the world consumption of milk. The importance of milk cannot be over-emphasised. Japanese are reported to have grown four inches taller over two generations ever since they started consuming more milk. China now wants to produce more milk so as to improve the quality of the national diet.

India has now emerged as the second largest producer of milk in the world, after United States of America and aims to become the largest producer by the year 2000, which is not all that far. Currently, the Indian per capita consumption of milk has nearly doubled over the past twenty years to 66 kg/year.

Growth of Milk Production and Milk Marketing

The Development of milk production in many parts of Asia is based on utilisation of crop residues and agricultural by-products. Milk production is thus fully integrated with the rest of agriculture. The crop residue-agricultural by-products, dairying system now yields as much economic return as the main crops like wheat, rice, millets, maize etc.

Milk production in Asian countries is demand driven (even at the cost of imports) on one hand and production driven (to utilize crop-residues) on the other. The income elasticity for expenditure on milk in India for example is 1.5. For every 1% increase in income, the expenditure on milk goes up by 1.5% of the increased income.

India has some 90 million farming families cultivating 140 million hectares and rearing some 90 million milch animals. With a cropping extensity of 1.3, the average cultivated land holding is 2 hectares of land and a milch cow or a buffalo. The income from the dairying system matches the income from the main crop.

Role of Buffaloes

Buffaloes play an important role in milk production in Asia

as they are fairly efficient converters of wheat/paddy straw into milk and produce 35% of milk in Asia.

Import Led Marketing

In several countries in Asia, imports of dairy products, both as finished products in consumer packs and dairy commodities for recombination into milk have helped in creating a market and a processing industry.

Problems and Constraints

Low productivity of cattle is the most important problem in the dairy sector in Asia. Though Asia has one third of the world cattle and buffaloes population it only produces 20% of the world throughput of milk. Milk production in Asia is also highly seasonal with the procurement of milk in India during the lean summer months dropping to as low as 30 – 40% of the flush season. Urban milk production is another curse on the dairy sector and this needs to be shifted to the rural areas as has happened in some parts of the world.

Poor infrastructure for transportation, processing and marketing of milk combined with high ambient temperatures is a challenge to the diary industry in many of the Asian countries. Poor quality milk received by the processing plants is a major constraint in producing world class products.

The availability of oil cakes (India is one of the largest exporters of oil cakes in the world) and crop residues is a major opportunity to increase milk production in the Asian region.

Dairying is a labour intensive activity and there is no shortage of manpower in many of the Asian countries which can be effectively used. Good land, plenty of water, high ambient temperatures and sunshine is another resource which cuts down on the capital expenditure needed on animal housing etc.

Regional Dairy Network for Asia

There are several Asian institutions engaged in regional

cooperation. The foremost institution that is quite active in ASEAN, Asian Development Bank and now the South Asian Association for Regional Cooperation (SAARC). These associations for regional cooperation aims at setting up preferential trading blocks and hopefully one day, common markets.

The SAARC council has recently sponsored a study to explore the possibilities of setting up dairy projects in the region. One of the major recommendations of the study is to have a SAARC Dairy Association and a SAARC Milk Grid. These measures would promote regional cooperation and draw on the natural advantages of the region. This would also lead to better utilisation of feed resources of the region and more efficient use of cattle and buffaloes for milk and meat purposes.

Institutions Involved in Dairying

Many countries in the Asian region have apex level organisations for the development of the dairy industry in the form of Milk Boards, Dairy Development Boards/ Corporations and Milk Producers Organisations. Many countries also have Dairy/Dairy Science Associations of professionals involved in dairying. Most countries in the region have now opened up milk processing to the private sector as part of the liberalisation of their economies. The private sector is thus making inroads into this sector which in many cases was the domain of the cooperatives/public sector. The role of Government is now seen as that of the guiding, promoting and monitoring authority.

Asian Dairy Network

There is need for an Asian Dairy Network which can help in optimal utilisation of regional resources. This includes resourceslike feeds, cattle, buffaloes, equipment, manufactures, training facilities, research capabilities and professional skills in planning, implementing and management of dairy development programmes and diary projects.

This would hopefully lead to more efficient milk production, processing and marketing of milk and milk products. Initially, the network can be started in SAARC and ASEAN regions and then integrated into an Asian Network. The Network can initially be an information exchange, receiving and disseminating real time information on milk production, stocks, prices future demand and availability of milk and milk products. It can also serve as an information pool for scientific research and technical information emanating out of the region and linked to other regions of the world.

Global Issues in Dairy Development

In subsistence agriculture the producer is also the consumer. There is no transport or processing involved and the consumers preference and taste is well known by the very same producer. In "rural economies" more than 50% of the population is involved directly in farming and the primary production of food and agricultural outputs. In the more expanding and diversifying economies the ratio of consumer to producer increases tremendously. The scenario here is that the employment in the food chain after farm gate is 5–10 times that in the primary production.

This development can be demonstrated with an example from Canada: In 1900 45% of Canada's population were employed on farm, whereas in 1990 only 3% were still directly involved in food production, but now 25% had found employment in various elements of the food chain. This is the general development seen in most industrialised countries, and it is rapidly under way in the developing countries, with a large variation in the scenarios seen. This change in the food supply system might be even more dramatic in the developing countries than during the industrialisation of the now developed countries because of a more rapid urbanisation. The balance is different and so are many of the problems and solutions.

The cities and towns of Africa are growing rapidly, and more rapidly than the rural populations. It is well known that the urban per capita demand for dairy products is higher than rural demand. Predictions for future demand for livestock products are staggering. Winrock estimate an increased demand from 1990 to 2010 for meat to 120% and for milk to 70%. Therefore, the near future will show a growing demand for dairy products and present both a challenge and an oppourtunity for development of the dairy sector in most developing countries; most-if not all- East African countries have the potential to produce enough milk to satisfy the domestic demand.

Many countries have seen a peri-urban sector develop very fast around or in the largest urban centres, responding immediately to the market demand and profiting from the lack of links between the rural producer and the urban consumer. Parallel to this development Tanzania has also seen private entrepreneurs initiate milk collection from Maasai pastoralists living in the vicinity of Dar es Salaam.

Moreover, projections of prevailing trends in food supply in Sub Saharan Africa is demonstrating an enormous increase in food deficit up to 8 times from 1990 to 2025. Interventions in the dairy sector should be market or demand driven and thus promoting a general economic development; FAO's emphasis is to enhance rural development through assistance to small scale dairy development within a mixed farming system with the objective to improve food security and to achieve sustainable development of agriculture.

Development of the dairy sector is an efficient tool in this context as it generates a continuous flow of income, diverse risk, improves utilisation of resources, and generates employment also outside the farming community because of the need for collection, transport, processing and marketing. Emphasis should be on the mixed farming system, where the animal component also increases the crop production. However, the pastoral production system also has potential for delivery of livestock products to urban centres depending on the distance to these centres but even more on the market orientation of the pastoral population. Two of the major general constraints should be mentioned here. The first is the lack of organisation and infrastructure. With the small amount of milk from each individual farmer some kind of common action has to be taken for the farmers to achieve the highest possible price for their milk. The more distant the market the more difficult and demanding is the marketing. The second major constraint for developing marketing based on small scale processing is the shortage of personnel trained and able to operate and manage these small units.

Why market orientation? For the farmer, it is a question of generation of income through utilisation of crop residues and other available resources for the development of the family. Income is necessary to take part and advantage of progress that would otherwise pass by - like for example schools; doctors; roads; water etc. For the society a market oriented agricultural production would secure food supply to the rapidly growing non-farming community; create employment and promote an economic development and provide import substitution or even products for export. Marketing services are critical to rural as well as urban food security. In the past we have seen many interventions for increasing the production, much less, however, in processing (except for investment projects) and minimal in marketing, transport and other supporting services essential for linking the producer and the consumer.

Urbanisation and Economic Growth

Urban populations in developing countries are expanding and expected to more than double by 2025. Moreover, by 2025 it is expected that *more than 50% of the world's population will be living in cities of more than one million people.* The exodus from rural to urban areas is most prominent in developing countries, as the industrialised countries have already reached this level of urbanisation through a continuous development over the last century.

The effects of this urbanisation are multiple. Rapidly-growing urban demand will be the major factor in shaping development of livestock production and marketing in the coming years. A significant increase in production, processing and marketing efficiency will be required to meet this rising demand from domestic resources. In the past, livestock development had not adequately taken the urban market into account and the interventions had not been market or demand driven but largely focused on production technology only.

No major studies have been carried out to ascertain what opportunities, challenges or negative impacts urbanisation presents for animal agriculture and associated milk and meat marketing nor for the possibility of food security and food self-sufficiency, which is a priority for most developing countries. Food production is already insufficient to satisfy the needs of most parts of the developing world. For example in Africa, projections indicate a deficit widening from 14 million tonnes of food in 1990 to 125 million tonnes in 2025.

Urban markets for food generally require a wide range of products, from expensive value-added foods to low cost commodities for lower income groups. Livestock products fit into the upper end of the spectrum and consumption of these products rises with income. However, the *high nutritive value of animal products* should not be overlooked as *they constitute an important source of vitamin, minerals and amino acids.* Animal products should be seen as an important supplement to complement the diet of poor people, which is frequently based on one or two food crops.

Only in few developing countries have the meat and dairy industries reached a state of developement to respond to the enormous challenge of providing safe, nutritious and regular supplies of livestock products to their rapidly expanding urban populations. Many developing countries have not been able to set up handling and processing facilities to cope with the fast pace of development at the urban level. Because they have not been able to convert local agricultural products into suitable foods for distribution in the cities, *the food industry has depended increasingly on imported materials or products.* Independent small-scale producers lack the economic strength to negotiate favourable terms for their business. Therefore, the means and ways to *support farmer's organisations* (societies, associations, cooperatives and unions as well as traditional community organisations) *will have to be carefully addressed* for the future benefit of farmers and the domestic livestock production.

In recent years we have, however, seen a reduced role of imports of dairy commodities; the outlook is that this reduction will continue in the future. There are a number of reasons for the change in the role of dairy imports. Most importantly, there is a reduction in surplus dairy products in Western Europe and USA, because of a change in policy and prices have increased. Food aid in the form of dairy commodities will not be as common and "dumping" of dairy products will occur more infrequently than in the past. A normalised exchange rate and the removal of subsidises in many developing countries has also increased the price on imported goods and made them relatively expensive. A growing interest in local milk production and to supply the market has become evident, although there is still a substantial import of dairy products in many African countries to satisfy the demand, mainly from the rapidly growing urban centres.

The high import has in the past encouraged governments (and donors) to invest in large scale milk production and industrial dairy plants around the major cities to supply the urban population with modern dairy products - Western style. These plants have, however, continued to depend on imported commodities for recombination and in most cases only a small part (in Tanzania less than 1 %) of the total dairy output was marketed through the large commercial processing plants operated by government parastatals. The provision of modern equipment to developing countries (without secured supply of spare parts and other services) has been a most wasteful feature of many development programmes. The system made little effort to promote an increased milk supply from rural small holders to the industrial plants. The traditional sector provided most of the total production. However, virtually nothing was channelled through the commercial factories but marketed and consumed within the local community. The traditional sector has, notwithstanding its major contribution to total milk production, been neglected. The rapidly growing demand in the expanding urban populations presents a great potential for development of the local dairy sector also because increased output will be absorbed by the market and not have a negative influence on the price

The rapidly growing urban populations in developing countries represent a strong demand for meat and milk now and in the future - *a demand which could be met by developing the domestic resources* in the majority of countries. This would benefit rural development and poverty alleviation through the creation of employment and income within the livestock sector and the service industry. Such an improvement of rural living conditions could possibly prevent, or at least reduce, migration of rural people to already overcrowded cities.

Demand and Market—the Link Between Producer and Consumer

It is the rapid urban growth and the even more rapidly growing demand for meat and milk that presents a particular challenge to the livestock sector and the dairy farmers, particularly in preservation and processing of the livestock food products. The future will therefore show a fast growing food industry, which will present the best - and so much needed - opportunity for employment in developing countries.

The agro-industry has in the past been located close to or even in urban areas. But the importance of efficient and reliable links out to the production areas have too often been overlooked. An alternative would of course be to locate the industry in a rural setting.

This would again require a good infra-structure, however, it would also provide a number of advantages such as: early preservation (perishability); integration of production and processing; employment in rural areas (discourage urbanisation); diversification - serving rural communities is much less complex than also serving urban centres. The location, however, is critical to viability and efficiency and should be based on the availability of necessary infra structure and on economic reasons rather than political preferences.

The benefits of well developed agro-industries can be listed as follows:

— Protect and preserve food for safe storage and transport from areas and seasons with surplus to those with a deficit;

— Transform animal raw materials into food products of a quality and quantity adequate to satisfy demand and nutritional needs; and into products which are acceptable, accessible and affordable;

— Transport and market fresh and processed products efficiently and economically.

The more distance in time and space between producer, processor and consumer the more critical is the need for safe and sound process and product control; for reliable systems of packaging and for a distribution network. There are serious hazards to human health from food which is inadequately preserved and processed. On the other hand the processing has also to be adapted to the need and requirements of the consumers. To present UHT milk to consumers who by custom boil the milk before consumption is not appropriate, but a result of central processing units prepared for supplying the urban centres and based on experiences from the industrialised countries.

There is now an urgent need for efficient and appropriate food and related agro-industries and distribution network more than ever before to link the domestic production with the dominating markets in the urban areas.

Future interventions would have to encourage a systematic approach to ensure farmers produce what will satisfy market demand. This includes consideration of transporting fresh produce from rural to urban markets to be delivered at a price and in a condition that is acceptable, accessible and affordable by urban consumers - with low and high incomes.

It is importantlt is important to integrate effectively production systems and the farmers with the industrial system that embody collection, preservation, processing, marketing and distribution together with the diversity of supplying services. Higher productivity in the primary production must be complemented and sustained by logistically and technologically efficient systems throughout the chain from farmer to consumer.

The strategies would have to relate to:

- The logistics of collection, transportation and distribution networks from rural producer to urban markets;
- Dispersion and location of facilities and centres for reception, processing and distribution; it is particularly important to prepare clear conclusions and recommendations concerning the differences between rural and urban located facilities;
- The need and opportunities for farmers to empower and organise themselves for participating in the food chain also outside or after the farm gate. It is important to recognise that value added through processing and marketing is more than double the value of the primary product;
- Cooperation between farmers organisation and the private entrepreneurs providing or offering input supply and support services.

Marketing provides many social and economic benefits, and only by participating will the producers fully utilise the opportunity for economic growth that is accessible in dairying.

Milk Producer Marketing Groups: A Cas Study in Uganda

Since the 1960s one of the most critical problems facing dairy farmers in Uganda has been recognised as that of marketing their milk. This problem has been recognised in the overall context of the importance of marketing considerations not only in stimulating increased milk production but also in raising dairy farm incomes and living standards and improving the nutritional well-being of the population in rural as well as urban areas. Hence, in Uganda the development of milk marketing infrastructure has been inextricably linked with the development of the dairy industry.

Although Government dominated the early initiatives in organised milk marketing in Uganda, a few independent producer marketing groups were established. Notable among the early groups established in the 1960s were Toro and Kigezi dairy co-operative societies.

However, all the development initiatives in the dairy sector got a setback in the 1970s on account of civil disturbances and political instability. It was not until 1987 that a serious programme to reconstruct the national economy was put in place. Accordingly, the Uganda Government prepared the National Rehabilitation and Development Plan for the period 1986 - 1990 which was later extended to 1992. The plan identified the Rehabilitation of the Dairy Industry as a priority programme whose overall goal was to regain self-sufficiency in milk through;-

- restoring production on dairy farms;
- improving milk marketing; and
- strengthening dairy extension services.

The Government programme was the basis for co-operation with many multilateral external donor agencies in the development of the dairy sector in Uganda. To a very large extent, implementation of the programme was co-ordinated by the UNDP funded and FAO executed technical assistance project, UNDP/FAO Dairy Industry Development Project UGA/84/023 and follow-on UNDP/FAO Rural Community Dairy Production and Marketing Project UGA/92/010.

Background

Agriculture dominates Uganda's economy, contributing about 60% of the Gross Domestic Product (GDP). The Livestock sub-sector contributes about 20% of agricultural GDP. Of the total area of 241,038 sq. km, 197,097 sq. km arc covered by land, most of which is arable. The country has good rainfall (annual rainfall Kampala (1993) 961mm) and low temperature variability (annual mean temperature, Kampala 22.0C).

The total cattle population is estimated to be 4.2 million [1] out of which 3.4 percent are improved dairy breed types (exotic and cross breeds). The main milk-shed areas are to the south of latitude one degree north extending from Mbale in the east to Kabarole in the west and Kabale in the south.

The milk production system range from semi-nomadic pastoralism at one end to zero grazing at the other end. In terms of feeding/management practices, three major systems of production can be identified as:-

1. Pastoral farms with large numbers of (greater than 50) indigenous stock grazing in coarse pasture throughout the year and milked twice a day. No supplementary feeding is provided.
2. Small scale crop and livestock farms close to urban centres using mixed breeds of dairy cows (less than 10).
3. Specialised dairy farms which keep grade or pure dairy breeds (20 - 100) largely on planted pastures, supplemental with commercial feed stuffs.

In 1990/91 total milk production was estimated to be 365 million2 litres per annum. However, more recent estimates made in 1993 put total milk production per annum at 553 million3 litres. Assuming that 40% of the milk is retained and consumed at farm level, 60% is marketable surplus. In Uganda only about 10% of the marketable surplus of milk is marketed through the formal (mainly Dairy Corporation) channel. Hence, producers have to use various informal and unreliable marketing channels to dispose off 90% of the marketable surplus. This is the real magnitude of the milk marketing problem in Uganda. It is the challenge which milk producer marketing groups are attempting to meet.

Development of Producer Marketing Groups in Uganda

Types of groups

In Uganda four types of milk producer marketing groups can be identified: -

(i) Producer marketing groups which operate as independent legal entities but sell a substantial portion of their surplus milk to the state-owned Dairy Corporation. Examples - Toro and Kigezi dairy co-operative societies.

(ii) Producer marketing groups which are completely dependant on Dairy Corporation for marketing their milk. Examples - Rubaare, Rugarama, Ibanda co-operative societies.

(iii) Producer marketing groups which operate completely outside the Dairy Corporation in milk surplus areas. Examples - Kamuli and Nakago co-operative societies.

(iv) Producer marketing groups operating in milk deficit areas but outside Dairy Corporation's milk collection system. Examples - Mbale and Lambuli dairy co-operative societies.

The period before 1987

Immediately after Uganda's independence in 1962 attempts were made to establish milk producer marketing groups. Accordingly, a few groups were established notable among which were Toro, Kamuli and Kigezi milk producer marketing groups.

One important feature in the formation of these early groups was that the initiatives for their formation came from organised milk markets. In the case of Toro Dairy Co-operative Society, it was the management of nearby Kilembe Mines Ltd which prompted the society's formation. Kilembe

Mines Ltd was a big copper mining complex in Western Uganda employing thousands of workers in the 1960s. A small pasteurisation plant with a capacity of 6,000 litres per day was installed by the society in Fort Portal with Kilembe Mines financial support. All the milk produced was brought by the mining company.

It was therefore not surprising that when the copper mines subsequently closed in the 1970s the society's operations also came to a standstill.

In the case of Kamuli producer marketing group the initiative for its formation came partly from the Jinja based Madhvani Group of Companies. When in the 1970s the Asian owners of the Madhvani Group were expelled from Uganda resulting in the closure of many of their operations, the small Kamuli plant was completely abandoned. Kigezi Dairy Co-operative Society fared a little better in that it was able to operate albeit poorly during the 1970s. The big difference between Kigezi Dairy

Co-operative society and the other groups was in the fact that the society was well patronised by its members and to a large extent owed its establishment to the dairy farmers themselves.

The period after 1987

The period after 1987 was marked by a serious effort by the Uganda Government to rehabilitate and develop the dairy industry. This was also the period that old producer marketing groups were revived and new groups established. All these activities were made possible by the implementation of two main dairy projects, viz, Government of Uganda/ UNDP/FAO Dairy Industry Development Project UGA/84/ 023 and GOU/UNDP/FAO Rural Community Dairy Production and Marketing Project UGA/92/010. Substantial support was received from World Food Programme, DANIDA, ADF, USAID and other donor agencies. By 1992 overall external donor commitment to Government Programme as a whole was US $ 55.1 million.

In 1986 there about 30 registered milk producer marketing groups out of which only one was operating. However operations of Toro, Kigezi and Kamuli societies were revived after 1987 with assistance from the project and seven other groups were established. These societies were:-Mbale, Lambuli, Kyagukuuju (Rubaare), Rugarama, Ibanda, Nakago and Mpigi dairy co-operative societies. Other societies and groups were also established during the same period although they did not work directly with the UNDP/FAO dairy projects. These groups included:- Masaka, Mityana and Iganga dairy farmers' associations. By 1993 there were over 50 registered dairy co-operatives, associations and companies handling about 1% of the total marketable surplus of milk.

Strategies adopted

Since milk-sheds in Uganda cover a vast area, and because of limited personnel and financial resources at the disposal of the dairy projects, the pilot scheme strategy was used to achieve objectives in the milk production and farmers' groups components of the UNDP/FAO projects. Ten farmers' groups were eventually selected. These groups selected represented different agro-ecological and farming systems throughout the various milk-sheds of the country, involving about 800 farms.

The primary objectives of promoting producer-marketing groups were:-

(a) increases in milk production per animal and per unit area at selected farms through appropriate technical interventions, training and demonstrations provided to farmers in organised groups.

(b) increased income for selected small scale and medium scale dairy farms through increased sale of surplus milk arising from organised marketing.

A model for field advisory (dairy extension) services and promotion of producer marketing groups was developed. It has now been adopted for use by the animal production department of the ministry of Agriculture, Animal Industry and Fisheries. This model is a package of integrated activities that includes:-

(i) Identification of dairy farming groups and carrying out of a baseline survey to identify constraints that stand on the way to increased milk production and improved marketing.

(ii) Group formation

(iii) Training, demonstration and advisory and other services provided at the farm level.

(iv) Technical assistance in milk marketing.

(v) Demonstration of small scale milk processing methods

(vi) Training in management of group projects

(vii) Provision of credit.

The model also includes a team of 5 field extension officers. The team is led by a Dairy Advisor and includes a veterinary officer, an artificial inseminator and two animal husbandry officers. The five officers however, also assist other types of farmers i.e. beef, poultry, pig, etc found in the same area. The team also continually co-opts other specialists from various departments who assist in conducting specific training courses and field demonstrations.

Identification of dairy farming groups

Where no producer-marketing group already exists in an area the first step is the characterisation of the milk production system in physical, biological and socio-economic terms. If this information is not readily available from documented literature, it is imperative that a quick baseline survey be undertaken before any intervention is undertaken. Such a diagnostic survey not only describes the system as it exists, but also helps identify training needs and the technologies and production inputs required to overcome the constraints.

The experience of the UNDP/FAO dairy projects in Uganda has been to select a group size of about 120 small and medium dairy farms scattered in an area of about 15 miles radius from a Dairy Corporation milk collecting centre or a major milk market. Ten such groups were identified.

Group formation

After the initial identification of a dairy farming group in an area, there is need to educate members of the group about the benefits of working. together. At the beginning members of the group may prefer to work together in some form of an informal or loose association. This is necessary because the farmers themselves must be convinced about the necessity of working together in order to solve their common problems.

It is also important that at this stage farmers arc exposed to the different forms of associations so that when it becomes necessary they can intelligently choose the type of association which is suitable to them.

Training, demonstration, advisory and other services provided at farm level

Depending on the constraints to milk production identified earlier, a series of relevant technical interventions need to be developed and implemented especially in feeding, breeding, disease control, management of dairy herds, regular veterinary and AI services to these farms. These interventions are carried out through training, demonstration and advisory services.

During the life of UNDP/FAO project in Uganda, these activities were co-ordinated by two regional extension advisors each working with five groups in south-western, central and eastern parts of the country. As mentioned earlier in this paper, each group was assisted by a team of five extension officers.

Non-residential training was often conducted at very low cost in locations within the group areas. Some demonstrations were carried out on members' farms while in some cases study visits to other areas were organised. A farm records system was introduced and farmers were taught to maintain these records as part of an integrated herd management scheme.

Technical assistance in milk marketing

Each group was provided with technical assistance in milk marketing. Each group was introduced to the advantages of collective milk marketing. They were encouraged to aim at satisfying demand in the neighbourhood towns and villages before delivering the surplus milk to more distant markets. They were taught the basics of running a milk collection centre. The groups which were managing milk collection centres were - Kamuli, Nakago (Nabuka), Toro and Kigezi co-operative societies.

One of the strategies suggested was for dairy co-operation societies to take over the running of the milk collecting centres presently owned and managed by Dairy Corporation.

Demonstrations of small scale milk processing methods

One of the ways of over-coming marketing problems facing the rural farmer in Uganda was identified as small scale processing of milk to improve quality and to develop other products for consumption and sale in the neighbourhood markets. At farm level, demonstrations were conducted to acquaint farmers with simple cost effective ways of producing ghee from surplus milk which could not be absorbed by the market.

At group level, the project introduced simple, low investment milk processing technologies on a trial basis at two (2) milk collecting centres; i.e. Mukono and Kamuli in co-operation with the dairy cooperative societies of the two areas. The trials were enthusiastically accepted by the groups.

The trials included pasteurisation (batch heat treatment) of milk using a charcoal stove and an electric boiler. The pasteurisation process using these systems resulted in milk of acceptable quality in terms of bacteriological count and composition.

The packaging of milk in plastic sachets using a simple electric hand sealer was also successfully demonstrated. For the societies, the packaging of milk in plastic sachets made handling and marketing of milk much easier and more hygienic. For the consumer, the packaging was also a much more hygienic presentation and provided insurance against adulteration.

The second series of trials carried out was on product development. These were carried out at the Kamuli Dairy in co-operative with the society. Products tried were cheese, butter, ghee and fermented milk. Pasta Filata and quasso blanco were cheese types tried and they both had good acceptance in larger towns and certain communities.

In view of the success of these activities, the project facilitated Kamuli milk collecting centre operated by Kamuli Dairy Co-operative Society to make it a training centre for small scale milk processing technology.

The project also proposed to Toro Dairy Co-operative Society to establish a mini-dairy plant as a long term solution to its marketing problems. The proposal was taken up by DANIDA who financed the establishment of a mini-dairy plant at a cost of US $ 250,000. The plant now processes pasteurised milk and produces cheese. This has had an enormous and positive impact on the operations of Toro Dairy Co-operative Society.

Training in the management of Group projects

The project realised that the ultimate success of the producer marketing groups depended on their ability to manage their group projects for the benefit of all the members. Groups implementing viable projects had a good chance of succeeding.

Training was therefore organised separately for all the members and for their leaders. Since all the groups working with the project opted to form co-operative societies, the first thrust of the training programme was in basic co-operative principles for the members and co-operative management for the leaders.

The second and equally important emphasis in the training programme was on managing specific income generating projects. Each group selected an income generating project which was financed through a small and medium scale dairy farmers credit scheme developed by the project and funded by the Dairy Development Committee. The different income generating projects selected by the various groups are shown in table 1.

Group leaders were not only introduced to general management principles and accounting but were given instructions on the specifics of what needed to be dune to ensure the financial viability of each project.

Provision of Credit

The project recognised the importance of credit to small and medium scale dairy farmers. Accordingly, a credit scheme was designed to meet the needs of these farmers who normally have very limited access to credit from financial institutions.

Table 1. Projects identified and credit financing requested by ten dairy co-operative societies

Dairy Co-operative Society	Project Identified	Credit Required
1. Mpigi	Pasture seed production and Farm input supply shop	15,000,000.00
2. Nakago	Financing of Building and infrastructual facilities for collective milk marketing.	26,000,000.00
3. Lambuli	Re-stocking of Dairy Farms	15,000,000.00
4. Kamuli	Transport facilities for Milk marketing	24,951,840.00
5. Toro	Transport facilities for Milk marketing	24,951,840.00
6. Rugarama	Operations of a Milk Cooling Plant at Rugarama	24,951,840.00
7. Kyabukuju	Operations of a Milk Cooling Plant at Rugarama	24,951,840.00
8. Kigezi	Farm Supply Shop	6,000,000.00
9. Ibanda	Farm Supply Shop	6,000,000.00
10. Mbale	Re-stocking of Dairy Farms	15,000,000.00

However, the project was also aware of previous credit schemes which failed mainly because of the numerous problems faced by borrowers and the lending institutions resulting in bureaucracies, high interest rates and a high rate of default in repaying loans.

The new credit scheme was funded by Dairy Development Committee with an initial fund of US $ 140,000. The first beneficiaries of the scheme were the ten dairy co-operative societies that had been working with the project.

The project in close consultation with the co-operative societies helped to develop viable projects for funding through the credit scheme. A list of projects identified and approved for funding arc shown on Table 1.

Producer marketing groups and government dairy sector policy

After the completion of the first phase of the rehabilitation the dairy industry in Uganda in 1992, Government decided to adopt a total dairy sector approach as opposed to interventions in limited areas. A complete review of official dairy sector policy was undertaken through a comprehensive dairy sector Master Plan study. The findings and recommendations of the dairy sector Master Pan

Study together with those based on UNDP/FAO project experience have formed the basis of a comprehensive Government dairy sector policy published in 1992. The official policy besides endorsing the UNDP/FAO Project model on producer marketing groups clearly emphasises the role these groups are expected to play in the future development of the sector.

Government emphasis in the promotion of producer-marketing groups is most evident in the policy on the privatisation of the state owned Dairy Corporation. According to official policy, privatisation of the Dairy Corporation will be undertaken gradually from the top as well as from the bottom of the Dairy Corporation structure. From the bottom, farmers' associations and co-operatives have already begun the process of purchasing or leasing Dairy Corporation cooling and collection centres. Hence, Dairy Corporation's operating milk collecting centres which by end of 1994 numbered 50 will each be owned and managed by a producer-marketing group, thus the number of producer marketing groups will double by 1996.

The second area of emphasis in the official dairy sector policy concerns the provision of support services to dairy farmers. According to Government dairy sector policy, Government support services will be rehabilitated and concentrated on milk-sheds selected on the basis of comparative advantage. Where the service benefits the dairy sector and the nation in general, the Government will fund the service. Where the service directly benefits individuals or groups and where they are willing and able to pay, e.g. AI services, the users will fund the service through user charges. Veterinary services will be privatised.

With regard to research, the policy clearly states that research will address practical problems faced by dairy farmers in Uganda and research priorities will be established in close co-operation with the farmers. The new dairy sector policy also includes the establishment of a Dairy Board to regulate, co-ordinate and promote development in the dairy sector. Farmers will be strongly represented on the Board. The policy also introduces a unified extension service.

References

Master Plan for the Dairy Sector, Ministry of Agriculture, Animal Industry and Fisheries, Uganda, 1993

Rural Community Dairy Production and Marketing Project UGA/92/010 TERMINAL REPORT GOU/FAO, 1993

Shafer, D., "Natural grazing Super-natural benefits: 'Modern prairie' boosts pasture production", *The New Farm*, May/June 1992.

Shirley, C., "Less milk, more profit: Organic feed and rotational grazing keep this dairy green", *The New Farm*, Sept/Oct. 1991.

Bibliography

Adams, Don et al., *Matching the cow with forage resources*, Rangelands, April, 1996.

Akin, D.E. & Chesson, A., "Lignification as the major factor limiting forage feeding value especially in warm conditions", *Proceedings of the XVI International Grassland Congress*, Nice, France, 1989.

Amble, V.N. & Jain, J.P., "Comparative performance of different grades of crossbred cows on military farms in India", *Journal of Dairy Science*, 1967.

Anon, "Breeding to improve milk production in the tropics", *Activities in Turrialba*, July 1973.

Arora, Dinesh ed, *Dictionary of Animal Husbandry*, Biotech Books, New Delhi, 2005.

Arora, Dinesh ed, *Dictionary of Dairy Science*, Biotech Books, New Delhi, 2006.

Arora, Dinesh ed, *Dictionary of Veterinary*, Biotech Books, New Delhi, 2004.

Atkinson et al, *Diseases of Cattle*, Biotech Books, New Delhi, 2008.

Avendano, L., "Reduction of heat stress in dairy cattle through the utilization of cooling systems", *Cuban Journal of Agricultural Science*, 1995.

Bailey, J.C. & Brace, A.W., "Experience with aluminium milk cans", *Journal of the Society of Dairy Technology*, 1956.

Barber F.W., "Hygienic aspects of the processing of butter, cheese, flavoured and fermented milks, and ice-cream", *Milk Hygiene*, WHO/FAO 1962.

Bernabucci, U. & Calamari, L., "Effect of heat stress on bovine milk yield and composition", *Proceedings of the Milk Yield and Quality in*

Cows Raised in Hot Environment Seminar, Zootecnica e Nutrizione Animale, 1998.

Bhat, P.N., "Kamaduk: the cow of plenty", *Indian Farming*, October 1972.

Branton., C., McDowell, R.E. & Brown, M.A., "Zebu-European crossbreeding as a basis of dairy cattle improvement in the U.S.A.", *Southern Cooperative Series*, Bull, 1966.

Carawan, R. E., *Processing Plant Waste Management Guidelines—Aquatic Fishery Products*, North Carolina Pollution Prevention Program, 1991.

Cecchini, G., "Keep it cool for cows", *African Farming and Food Processing*, 1998.

Chalong-Wichiraphakorn, "Methods of reducing heat stress in dairy cows", *Khon Kaen Agriculture Journal*, 1995.

Cooper, M M, *Beef Production*, Biotech Books, New Delhi, 2005.

Dado, R.G. & Allen, M.S., "Variation in and relationship among feeding, chewing and drinking variables for lactating dairy cows", *Journal of Dairy Science*, 1994.

Dawra, Sudhir, *Bird Flu: Diagnosis and Treatment*, Biotech Books, New Delhi, 2006.

Devendra, C, *Sustainable Animal Production from Small Farm Systems in Sou*, Daya Publishing House, New Delhi, 1998.

E B J, *Cows and Poultry in India: Their Care and Management 2nd edn*, Biotech Books, New Delhi, 2004.

Eckles, Clarence H & Ernest L Anthony, *Dairy Cattle and Milk Production*, Biotech Books, New Delhi, 2007.

Ellis, J.G., "Aseptic processing and packaging of milk", *Journal of the Society of Dairy Technology*, 1977.

Elye, L C R Norris, *Poultry Science: The Selection Rearing and General Treatment*, Biotech Books, New Delhi, 2005.

Fahey, G.C. Jr. & Hussein, H.S. "Forty years of forage quality research: accomplishments and impact from an animal nutrition perspective", *Crop Science*, 1999.

Falk, Dean E., and Robert M. Ohlensehlen, "Dairy Waste Management System Planning –*Estimating Storage*", *Cooperative Extension, Bulletin No.* 694, University of Idaho, 1989.

FAO, *Good Practices for the Meat Industry*, Daya Publishing House, New Delhi, 2007.

FAO, *Veterinary Education*, Daya Publishing House, New Delhi, 1998.

Ghosh, Bibek ed, *Crops and Livestock Farming*, Genetech Books, New Delhi, 2007.

Ghosh, Bibek ed, *Integrating Crops and Livestock*, Genetech Books, New Delhi, 2007.

Ghosh, Bibek ed, *Trends and Issues of Livestock Production Systems*, Genetech Books, New Delhi, 2007.

Godbole, N N, *Milk: The Most Perfect Food*, Biotech Books, New Delhi, 2007.

Gupta, P.R., ed., *Dairy India 1983*, New Delhi, Rekha Printers Pvt. Ltd., 1983.

Hall, H.S., "Retail packaging of liquid milk", *Journal of the Society of Dairy Technology*, 1973.

Hambidge, Gove ed, *Diseases and Parasites of Cattle*, Biotech Books, New Delhi, 2008.

Hambidge, Gove ed, *Diseases and Parasites of Dogs and Cats*, Biotech Books, New Delhi, 2004.

Hambidge, Gove ed, *Diseases and Parasites of Horses and Mules*, Biotech Books, New Delhi, 2008.

Hambidge, Gove ed, *Diseases and Parasites of Poultry*, Biotech Books, New Delhi, 2008.

Hambidge, Gove ed, *Diseases and Parasites of Sheep and Goats*, Biotech Books, New Delhi, 2004.

Hambidge, Gove ed, *Diseases and Parasites of Swine*, Biotech Books, New Delhi, 2004.

Hansen, Brandon D., "An Economic Model for Analyzing Alternative Dairy Waste Handling Systems," M.A. thesis, Department of Agricultural Economics, Washington State University, Pullman, December 1993.

Harper, W. F. and Blaisdell, J.L., "State of the art of dairy food plant wastes and waste treatment", *Proceedings of the Second National Symposium on Food Processing Wastes*, Denver (Colorado), USA, 23–26 March, 1971.

Harvey, V' M Clunie & Harry Hill, *Milk Products 2nd edn*, Biotech Books, New Delhi, 2006.

Heinrichs, A.J., "Dry cow management", In: *Dairy Management Manual*, A Western Regional Extension Publication, 1990.

Jay, J.M., *Modern Food Microbiology*, Fourth Edition, New York: Chapman & Hall, 1992.

John C.K., "Milk control on reception", *Milk Hygiene*, WHO/FAO 1962.

Jull, Morley Allan, *Successful Poultry Management 2nd edn*, Biotech Books, New Delhi, 2007.

Korten, D.C., "Rural development programming: the learning process approach", Rural Development Committee, *Rural Development Participation Review*, Ithaca, NY, USA, Cornell University, 1981.

Kumar, Ashok, *Rabbit Wool Industry*, Daya Publishing House, New Delhi, 2004.

Kutty, C Ibraheem & Sheeba Khamer, *Milk Production and Processing,* Daya Publishing House, New Delhi, 2004.

Kyle, Edwin Jackson et al, *Fundamentals of Farming and Farmlife*, Biotech Books, New Delhi, 1999.

Lander, P E, *Feeding of Farm Animals in India*, Biotech Books, New Delhi, 2001.

Lasater, L., *The Lasater philosophy of cattle raising*, Texas Western Press, 1972.

Lewer, S H, *Wright's Encyclopaedia of Poultry Science*, Biotech Books, New Delhi, 2000.

Littlewood, R.W., "Crossbreeding for milk", *Indian Journal of Vetenery Science*, 1933.

Madalena, F.E., "Crossbreeding strategies for dairy cattle in Brazil", *World Anim Rev*, 1981.

Marwaha, Kavita ed, *Meat Hygiene*, Genetech Books, New Delhi, 2007.

Master Plan for the Dairy Sector, Ministry of Agriculture, Animal Industry and Fisheries, Uganda, 1993

McGee, Harold, "Milk and Dairy Products", *On Food and Cooking: The Science and Lore of the Kitchen* (in English), New York: Charles Scribner's Sons, 1984.

Merges, George, and Slade, Roger, "Dairy Development and Milk Cooperatives: The Effects of a Dairy Project in India", *World Bank Discussion Paper*, Washington, D.C.: World Bank, 1987.

Meyn, K. & Wilkins, J., "Breeding for milk in Kenya, with particular reference to the Sahiwal stud", *Wld Animal Reverence (FAO)*, 1974.

Mielke, T., ed., *Oil world annual*, Hamburg, Germany, ISTA Mielke GmbH, 1993.

MILK INDUSTRY FOUNDATION, *Analysis of Milk and Its Products: A Lab Manual 2nd edn*, Biotech Books, New Delhi, 2005.

Morgan, Russell M., and Luther H. Keller, "Economic Comparisons of Alternative Waste Management Systems on Tennessee Dairy Farms", *Bulletin 656*, University of Tennessee Agricultural Experiment Station, Knoxville, 1987.

Mowat, Noel & Mark Rweyemamu, *Vaccine Manual: The Production and Quality Control of Veteri*, Daya Publishing House, New Delhi, 1999.

Mowat, Noel ed, *Quality Control of Veterinary Vaccines in Developing Countri*, Daya Publishing House, New Delhi, 1998.

Newton Clark R., "Water supply on the dairy farm, (English, French, Spanish)", *Milk Hygiene*, WHO/FAO 1962.

Owen, W Powell, *Poultry Farming and Keeping*, Biotech Books, New Delhi, 2007.

Pampori, Zahoor A, *Field Cum Laboratory Procedures in Animal Health Care*, Daya Publishing House, New Delhi, 2003.

Pegler, Henry Stephen Holmes, *Goats ànd their Profitable Management 9th edn*, Biotech Books, New Delhi, 2005.

Potter, N.N., J. H. Hotchkiss, *Food Science*, Fifth Edition, New York: Chapman & Hall, 1995.

Poulsen, P, Ronkilde, "Selecting packaging systems", *International Dairy Federation Annual Bulletin Document*, 1976.

Preston, T R, *Tropical Animal Feeding: A Manual for Research Workers/ FAO*, Daya Publishing House, New Delhi, 1998.

Rao, V M, *Evaluation of Women Dairy Project in Rajasthan*, Regency Publications, New Delhi, 2003.

Richmond, H Droop, *Laboratory Manual of Dairy Analysis 3rd Revised edn*, Biotech Books, New Delhi, 2004.

Robinson, R. K., *Modern Dairy Technology, Volume 1*. Elsevier Applied Science Publishers, London, 1986.

Rural Community Dairy Production and Marketing Project UGA/92/010 TERMINAL REPORT GOU/FAO, 1993

Rynk, Robert, *On-Farm Composting Handbook*, Northeast Regional Agricultural Engineering Service. Ithaca, NY, 1992.

Shafer, D., "Natural grazing Super-natural benefits: 'Modern prairie' boosts pasture production", *The New Farm*, May/June 1992.

Shirley, C., "Less milk, more profit: Organic feed and rotational grazing keep this dairy green", *The New Farm*, Sept/Oct. 1991.

Singh, Vir & Babita Bohra, *Dairy Farming in Mountain Areas*, Daya Publishing House, New Delhi, 2006.

Sinha, B.N., "Taylor cows of Patna", *Indian vetenery Journal*, 1951.

Sinha, S N, *Cow Keeping in India: A Simple and Practical Book on their C*, Biotech Books, New Delhi, 2008.

Smith, R.D., L.E. Chase and C.J. Sniffen, "Managing the dry cow for top production, reproduction and health", *In: Dairy Management Manual*, A Western Regional Extension Publication, 1990.

Speedy, Andrew & Rene Sansoucy eds, *Feeding Dairy Cows in the Tropics/FAO*, Daya Publishing House, New Delhi, 1998.

Srivastava, Manish L, *Handbook of Milk Microbiology*, Daya Publishing House, New Delhi, 2002.

Swasigood, H.E., "Characteristics of Edible Fluids of Animal Origin: Milk", In *Food Chemistry*, Second Edition, Revised and Expanded, O.R. Fennema, Ed. New York: Marcell Dekker, Inc., 1985.

Varnam, A. H. and Sutherland, J. P., *Milk and Milk Products: Technology, Chemistry and Microbiology*, Chapman and Hall. London, 1994.

Verma, Rishendra, *Glimpses of Indian Veterinary Science*, Daya Publishing House, New Delhi, 2000.

Wijeratne, W.V.S., "Crossbreeding Sinhala cattle with Jersey and Friesian in Ceylon", *Animal Production*, 1970.

Zuvakis, C., Jr., *Economic development: an introduction*, New York, NY, USA, St Martin's Press, 1979.

Index